STRATEGY AND HISTORY

STRATEGY AND HISTORY

Collected Essays, Volume 2

EDWARD N. LUTTWAK

Transaction Books
New Brunswick (U.S.A.) and Oxford (U.K.)

Library of Congress Catalog Number: 85-1032
ISBN: 0-88738-065-4 (cloth)
Printed in the United States of America

Library of Congress Cataloging in Publication Data

Luttwak, Edward.
  Strategy and history.

  (Collected essays/Edward N. Luttwak; v. 2)
  1. Military art and science—History—20th century—
Addresses, essays, lectures. 2. Deterrence (Strategy)—
Addresses, essays, lectures. 3. Sea-power—Addresses,
essays, lectures. 4. World politics—1975-1985—
Addresses, essays, lectures. I. Title. II. Series:
Luttwak, Edward. Collected essays; v. 2.
U42.L88   1985   355′.02   85-1032
ISBN 0-88738-065-4

To my son,
Joseph Emmanuel

Contents

Acknowledgments

The author gratefully acknowledges the following publishers and publications for permission to use previously published materials:

"The Problem of Extending Deterrence," in *The Future of Strategic Deterrence*, Pt. 1 *Adelphi Papers*, No. 160 (London: IISS, 1980).

"The Nuclear Alternatives," in Ed. Kenneth A. Myers *NATO the Next Thirty Years* (Boulder: Westview, 1980).

"Military Balance and Deus Ex Missiles," originally appeared as: "The Lebanon War and the Military Balance: Are Soviet Weapons Inferior?" in *Middle East Insight*, Vol. 2, No. 6 (March/April 1983), and "Deus Ex Missiles," in *The New Republic* (9 August 1982).

"The Political Uses of Sea Power: The Theory of Suasion," in E.N. Luttwak, *The Political Uses of Sea Power*, Studies in International Affairs No. 23 (Baltimore: Johns Hopkins University Press, 1974), pp. 1-38.

"Sea Power in the Mediterranean," in E.N. Luttwak, *Sea Power in the Mediterranean: Political Utility and Military Constraints*, The Washington Papers, Vol. 6, No. 61 (Beverly Hills: Sage Publications, 1979).

"War, Strategy, and Maritime Power," *Naval War College Review* (February 1979).

"The American Style of Warfare," *Survival*, Vol. 21, No. 2 (March/April 1979): 19-22.

"The Operational Level of War," *International Security*, Vol. 5, No. 3 (Winter 1980/81): 61-79.

"Low-Intensity Warfare," *Parameters*, Vol. 13, No. 4 (December 1983): 11-18.

"Strategies of the Nuclear Age," originally published as: (Part I) "Of Bombs and Men," in *Commentary* (August 1983); (Part II) "The

Sinuous Diplomatic Dancer," in *TLS* (15 October 1982); (Part III) "A Record of Failure," in *Commentary* (June 1983).

"Delusions of Soviet Weakness," *Commentary* (January 1985): 32-38.

Introduction

Strategy has become fashionable in the United States, though only the demand for that elusive commodity is in evidence, there being no perceptible supply. It was Vietnam that did it of course, by proving that good intentions and the sheer outpouring of means need not suffice to achieve even the half-victory that was desired. At least dimly, it was then recognized that the diverse instruments of statecraft ought to be harnessed to a hierarchy of coherent purposes within a definite grand strategy in order to obtain worthy results. And it was even realized, if only more dimly, that the conduct of war cannot be left to the natural proclivity of each military service and branch to maximize its own production, fighter-bomber sorties, patrol "contacts," body-counts, weight of shell or whatever, but should instead be controlled by the sharp choices of a definite military strategy.

After these discoveries were made, it became routine for opposition spokesmen to protest that America was adrift without a (grand) strategy, while those holding office, often enough frustrated in their own intragovernmental strivings for a minimum of coherence, could only feebly claim the distinction of a strategy for the assorted goings-on over which they presided. For neither with President Carter nor with President Reagan did the electorate obtain an administration actually willing to accept the severe discipline of a grand strategy, which must entail the ruthless suppression of interdepartmental rivalries and the irritation of large domestic constituencies in order to pursue consistent purposes to their logical conclusion.

Were the electorate's hopes betrayed? Hardly. The record of recent years shows quite conclusively that there was no true desire for the discipline and consistency of a grand strategy. It was the style, not the substance, of each president that was in demand at least for a while: Jimmy Carter's devolutionist stance was well received at first but promptly rejected when it came to deeds—even the Korean garrison found its congressional defenders clearly predominant when the attempt was made in 1977-78 to withdraw the troops. Similarily, Ronald Reagan's talk of "standing tall" was very popular, but so was his avoidance of any combat larger than the invasion of Grenada; there was scarcely any public complaint when Syria was left unpunished after sponsoring, protecting, and praising the terrorism that killed so many Americans both military and diplomatic.

In other words, the essential precondition of any grand strategy seriously meant, namely that should indeed be seriously meant, remained unsatisfied: The American public welcomes the intimation that a grand strategy is at work but does not desire its actual implementation if it entails significant costs and risks—as any grand strategy must. Rather than accept sacrifice, it is the greater control over world events that a grand strategy could yield that is willingly sacrificed.

A professional dedication to the craft of strategy could at this point evoke bitter complaint or inspire yet more energetic attempts to educate the public. But such reactions would collide with a fundamental aspect of the American condition, newly revealed in the post–post-Vietnam years, which no strategist can challenge qua strategist, and which the present writer for one has no desire to change: Absent the most urgent threat physically manifest, Americans insist on the priority of individual goals over the collective pursuit of positive state goals, in the realm of world politics as in any other; and without positive state goals of a sharply defined authority, a grand strategy worthy of the name cannot even be formulated, let alone pursued. If the ship has no set destination there is no point in sailing any faster.

The impossibility and indeed illegitimacy of imposing positive strategic goals upon a pluralist society not immediately threatened, does not prevent the United States from accumulating power and neither does it obviate all occasions for its use. But it does very largely limit the scope of American action in world affairs to a defensive reflex in protection of the status quo, for which there is no great need of a grand strategy.

There remains the lesser strategy comprised within the realm of military power itself, whose aim is to make a purposeful instrument out of the diverse forces provided by the diverse military bureaucracies. There is no cultural or constitutional obstacle to preclude such military strategy, yet we do not have it in a form tolerably coherent because of an institutional structure poorly made by obsolete legislation, which new legislation could reshape. At present, we attempt to convert money into military power through the medium of vast service bureaucracies, each dedicated to the perpetuation of its own preferred forms of military power. Aside from the feeble committees of the Joint Chiefs of Staff manned by officers each appointed by his own service to promote its own narrow interests, there is no national military authority; and no committee of coequal service representatives can go beyond lowest common denominator compromises to make choices and set priorities. With no mechanism to reduce less useful forces to make way for those more greatly needed, the priorities of the past are simply perpetuated in large degree, with no dynamic process of creative destruction to make way for the new. Genuine innovation to create new forms of military power is most reluctant, even as there is a very great eagerness to provide new and more modern equipment for the old forms of military power: the aircraft carriers persist in the missile age with new vessels of unprecedented size, while the navy builds none of the new hull forms devised by its own engineers, and at the national level, naval power is still assigned a higher priority than the land forces, as if it were still a maritime Japanese empire that we had to confront and not a land empire par excellence.

When there is any fighting to be planned and commanded, the same institutions which can do no more than share out the loot of the defense budget proceed to act out their structure by sharing out the fighting. Hence the multiservice circuses that are unleashed when the United States attempts combat even on the smallest scale, yielding tragedy and humiliation in the Iran rescue attempt, and farcical confusion in the seizure of Grenada. It is the characterizing feature of military strategy at all levels that it rejects compromise and imposes sharp choices (one can break through a front on the right, the left, or the middle, but not by diluting one's forces all along that front), but our present military institutions are capable only of feeble compromise. If it is part of the American condition that no grand strategy is allowed, that is certainly not the case for military strategy—and it is time to act, to achieve changes long overdue.

In what follows, the essential unity of strategy emerges from the study of seemingly very diverse questions. The reader will recognize a unity of method in the combination of the "vertical" analysis of the levels of strategy with the "horizontal" analysis in which adversaries and third parties are considered at each remove. The reader will also recognize a consistent bias, for strategy is not a neutral pursuit and its only purpose is to strengthen one's own side in the contention of the nations.

Part I

Deterrence, Destruction, and Detente

1

From Mutual Assured Destruction to Mutual Assured Security

"Assured Destruction" in Retrospect

Since any instrument of military power that *can* inflict damage upon an adversary (and whose existence is known) *may* deter him, a deterrent effect is latent in any overt military deployment. Until the emergence of nuclear weapons, however, the continuous deterrence generated by military forces was generally treated as a bonus gain. And this was a gain often too small to warrant prior disclosure of weapons whose existence could otherwise be kept secret, so as to enhance their surprise value in combat. Deployments were thus evaluated in terms of their combat capability alone, and General Staffs did not ordinarily feel obliged to evaluate the worth, and limits, of any putative deterrent effects.

When nuclear forces were first deployed by the United States in the late 1940s, explicit recognition was given to their deterrent effect, but their prospective combat use was also regarded as important, indeed possibly more so. It was consistent with this view that the deployment of continental air defenses should have followed that of the SAC manned bomber force, and if the salient output of the latter was primarily deterrent, that of the former was certainly a *combat* output.[1]

Whether by intent or tradition, a balance was therefore maintained during the 1950s between the ability to deter and defend, on the one hand, and the ability to attack and resist counterattack, on the other. All four were recognized as valid—strategically and politically. In particular, insofar as American strategic-nuclear power was harnessed to defend Western Europe, or rather to counter the political impact of the Soviet Union's (nonnuclear) preponderance on the ground, the ability to attack with nuclear weapons and to resist Russian retaliation upon the United States was essential to the American strategic arsenal.

3

Notwithstanding the justified criticism of the official doctrine of Massive Retaliation,[2] whose formal enunciation only came after the doctrine had seen its best years and was about to be invalidated, under the aegis of this strategy American strategic power performed two separate missions: (1) the simple deterrence of direct (nuclear) attack upon the United States and its allies and (2) the projection of American strategic-nuclear power in the political arena where it could compete effectively with (nonnuclear) Soviet military power.

What was criticized in the strategic literature associated with the concept of "limited war" was the attempt to extend the application of strategic-nuclear power beyond the limits set by its credibility-in-use. Even so, the critics did not go so far as to claim that these credibility limits coincided with the boundaries of the United States: they did not dispute the "convertibility" of strategic nuclear power to political purposes, so long as the interests to be promoted (e.g., the status of Berlin) were truly central, and recognized as such by all concerned.[3]

The great innovation of the Assured Destruction doctrine and of the force deployment policies derived from it was its unidimensional nature; instead of the four dimensions of military power—deterrence, defense, attack, and postattack defense (i.e., resistance to counterattack)—exclusive reliance was to be placed on a single dimension, deterrence. Now accorded the weight of a philosophical doctrine of general validity, the deterrence-only concept[4] was in fact an ad hoc response to a transitory combination of technological circumstances as well as an answer to an internal and "bureaucratic" political problem.

The emergence of ballistic missile technology during the later fifties came at a time when the development of the appropriate counter-weapon, a ballistic missile defense (BMD), was still embryonic. A time-lag in the defense response to an innovation in the offense (and vice versa) is normal, but in this instance it was in fact widely believed, in the United States if not in the Soviet Union, that there was no effective counter to the ballistic missile on the policy horizon. And when the absolute feasibility of BMD forces did become apparent, their effectiveness began to be evaluated in economic terms, and deemed to be wholly inadequate.

But the deterrence-only concept that formed the basis of the Assured Destruction doctrine was predicated on the superior merit of a system of deterrence that relied exclusively on survivable offensive forces. In this scheme of things, population defenses, antibomber as much as BMDs, merely reduced the reliability, or increased the force-level requirements, of the offense. It followed that adherence to the doctrine implied opposition to population-defense deployments on principle, and regardless of their costs, unless and until the latter would decrease (for a "fully" effective defense) to the point that the doctrine itself with its deterrent core, was abandoned.

The internal bureaucratic and political problem to which the doctrine was addressed was the persistent pressure emanating from the services and their congressional supporters for the deployment of additional strategic forces. Having come into office in 1961 committed to a substantial increase in American military power in all its dimensions, and having capitalized on the electoral value of the Missile Gap issue in 1960, the new administration did not maintain a common front with the advocates of increased "preparedness" for very long. First, the intelligence projections associated with the Missile Gap episode were invalidated; even before Inauguration Day, it became apparent that the United States in fact retained a net superiorty in offense strategic-nuclear forces, a superiority that would increase by a wide margin in the near future.[5] The new administration did approve increases in strategic forces, but the emphasis soon shifted from the urgency of additional deployments to a search for ways of utilizing the power thus made available. Second, the new administration was now constrained by the fiscal responsibilities of office, and its leader could no longer support the weapon programs promoted by the services without worrying about their cost. Third, men were active in the new administration whose greatest achievement and greatest challenge had been the abortive military policy associated with NCS-68.[6] This policy had been predicated on the limitations of nuclear armaments vis-à-vis indirect and low-intensity threats, and had called for the expansion of nonnuclear forces, especially ground forces. And the threats to which the new administration was increasingly oriented were precisely such low-intensity threats, in the Third World.

Before the final retreat to Assured Destruction was made in the later 1960s, the search for ways of applying American strategic superiority to political purpose was pursued diligently by the Secretary of Defense and his associates, men whose inclinations and disposition[7] did not, however, favor this search. The subtle and variegated realm of international politics does not lend itself to qualification and comparative numerical evaluation. In their search for the stable certainty of reliable systems, the systems analysts could only filter out the unquantifiable.

To the very end, Secretary McNamara did not attempt to totally deny the political convertibility of strategic nuclear power, even as between the superpowers, but the policy he supervised resulted in such a denial. The problem of the political applicability of strategic power was salient at this time for three reasons:

- The continuing perceived weakness of NATO vis-à-vis a Soviet Union that was still active, if more than a little erratic, in its revisionism—especially with respect to the status of Berlin and the wider German question.
- The challenge posed by the French nuclear weapon program, whose explicit rationale was the inadequacy of American deterrence on behalf of Europe. This inadequacy was obviously and exclusively *political;* the French did not dispute the capability of the American strategic arsenal, but doubted the intensity and consistency of the will to use on behalf of Europe and France.
- The very availability of the resource, abundant strategic power.

The fundamental difficulty that faced U.S. defense planners was that the level of American strategic power was then at an intermediate point within what was more and more explicitly seen as a discrete two-level system. Implicitly thinking exclusively in terms of *military* utility, the Secretary of Defense and his associates saw any level of strategic power beyond that required to kill, reliably, a satisfactory percentage of Soviets (and Chinese, and so on) as quite useless—unless and until the next recognized technical gradation, a "disarming first strike" capability, was attained. And this was a capability that could only be attained at very great cost, it was thought, and at the still greater risk of provoking an anticipatory attack.

Before the search was finally abandoned, the attempt was nevertheless made to utilize this intermediate margin of power that was redundant for simple deterrence and insufficient for preemption. The result was the short-lived Counterforce strategy unveiled in the June 1962 Ann Arbor speech. Counterforce was to be applied to the deterrence of Russian offensive moves that were of high intensity (i.e., an invasion attempt in Europe) but stopped short of a direct, nuclear attack. In concrete terms, it required that a portion of the offensive arsenal be targeted on Soviet military targets (including strategic forces) while the rest remained targeted on Soviet cities, so as to deter Soviet reprisals against American cities in the aftermath of a counterforce strike.

Perhaps because of the Cuba Missile Crisis which resulted first in incomprehension,[8] then euphoria, and finally in acute doubts over the overall utility of strategic power, there was a rapid erosion of confidence in the merits of the Counterforce strategy. On the one hand, the strategy had failed to solve the problem of European nuclear defense. In spite of Counterforce, the Europeans continued to manifest doubt of the validity of American "extended" deterrence, and the French continued to develop their *force de frappe,* which failed to satisfy the exacting survivability requirements specified by McNamara's systems analysts and which, therefore, could only be "useless and dangerous." The political value of this symbol of autonomy and national revival, its prestige value in the diplomatic arena, and the economic side-benefits of this technologically exacting endeavor, were all seemingly discounted by many American observers. On the other hand, if Counterforce was disappointing on the inter-Alliance political front, it was soon perceived as downright dangerous ("destabilizing," to use a favored word of those years) on the strategic front, vis-à-vis the Soviet Union. In the presence of the vast superiority of American strategic offensive power, the first-strike implications of Counterforce were apparent and very disturbing. Even when the Counterforce strategy was abandoned, the final retreat to the simple counter-city deterrence of the Assured Destruction doctrine was not complete.

Under the aegis of Damage Limitation, a diluted version of Counterforce[9] was to be pursued along with the prospective deployment of ballistic missile defenses. Phased-array radars and high-acceleration terminal interceptors distinctly improved the technical performance of the available BMD technology. If it remained cheaper to kill than to defend, the merits of the two activities were also not to be evaluated equally. There is no need to review in detail the evolution of the successive BMD projects promoted during the late fifties and sixties, from *Nike-Zeus* with its mechanically scanned radars, to the two-phase intercept, and overlapping phased-array radars of the systems that were developed by the later sixties. In the end, for all the evaluation of cost-exchange ratios, of the relative merits of "thick" and "thin" population defenses, and of terminal force defenses (which met wider approval), the rejection of the BMD option until 1967 was motivated by doctrinal consideration—i.e., its inconsistency with the tenets of Assured Destruction.

In terms of domestic and bureaucratic politics, the great virtue of the Assured Destruction doctrine was that it set a ceiling on the level of strategic armament that was necessary, or even useful. Once it was agreed that a deterrent threat, and that alone, could best ensure the security of the United States, and that no additional (political) purposes could be achieved by strategic-nuclear power, population-defense BMD forces could be declared redundant, (or destablizing or—inconsistently—both), and a ceiling could be set on the quantity and quality of the offensive forces required.

At first, the force-deployment rationales that could be derived from the doctrine were seen as useful in determining the minimum requirements of deterrence in terms of offensive power. Generous allowances for uncertainty were therefore in order:

- First, a level of damage to the population and industry of the Soviet Union that could be deemed to be "unacceptable" was to be defined;[10]
- Second, the number and type of weapons, reliably delivered on target, required to ensure this level of damage was easily, if arbitrarily, computed;
- Third, additonal weapons were added to precompensate for expected technical errors in weapons performance, operational control, or target allocation;
- Fourth, the entire force was to be diversified (in fact triplicated) in order to ensure its resilience in the event of a sudden technological breakthrough that would impair the survival or penetration capability of any one mode of delivery;
- Fifth, any additional force-level margin required by the penetration and postattack survival criteria[11] (as against the projected defensive and counterforce capability of the adversary) was to be provided.

In fact, of course, these were ex post facto rationales for weapons already deployed for simpler competitive reasons. Once a reliable deterrent was thus achieved vis-à-vis the Soviet Union, by extension, and without any attempt at consistent calculation,[12] the retaliatory threat was also held to be valid as a deterrent to less-than-total attacks (reasonably enough) and, much less reasonably, it was also held valid to deter (nuclear) attacks emanating from any and all adversaries. As a result, the force deployment criteria that could be derived from the Assured Destruction doctrine, with all the wealth of mathematical data that went with them in the way of visual aids, proved to be very useful to the administration in its successful resistance to pressures for the deployment of weapons such as the B-70 and its equally abortive successor, the RS-70.

The doctrine did not, however, suffice in itself to negate the value of forces other than those directed at simple deterrence. For this to be so, a further conceptual element had to be added, an element that would make the doctrine *reciprocal*. Mutual Assured Destruction (MAD) was based on a simple if heroic assumption, an equation of the strategic goals, values, and perspectives of the leaders of the Soviet Union with those of their American counterparts. If *both* sides saw their central goal as the deterrence of direct attack from the other upon themselves, and no more than that, it followed logically that both sides would maintain Assured Destruction forces, and no more than those. Since, in any event, belief in the political applicability of strategic power was waning in the United States, the deployment of additional offensive weapons oriented towards counterforce missions could be opposed on the grounds that they would undermine the stability of an Assured Destruction deterrent system that had now become a "mutual." Such opposition was much facilitated by the persistent failure of the promoters of counterforce-oriented forces to provide satisfactory *political* rationales for the deployment of forces over and above those required for simpler deterrence. At the same time, and for the same reason, the deployment of population-defense BMD forces could also be rejected: if a BMD worked it would impair the stability of Mutual Assured Destruction; if it did not, it would be useless. Moreover, a population BMD could be represented as both useless and dangerous, if it was assumed that the Soviet Union would take a prudent (i.e., pessimistic) view of its capabilities, and overreact accordingly.

Once Assured Destruction became a doctrine of reciprocal validity in the minds of American policymakers, it alone sufficed to circumscribe the scope of American strategic deployment policies. This aspect of the doctrine thus became a major asset in the administration's attempt to control defense expenditures once the war in Southeast Asia intensified. For the arms control interest too, the doctrine was a useful political device, though in their hands the doctrine underwent a process of erosion towards small-scale (or "finite," or "minimum") deterrence.[13] More significant still was the parallel erosion of the weapon-survivability criteria derived from the doctrine (but common to all stable systems of deterrence). This process reached a climax in the proposals made in 1969 for a "launch-on-warning" ICBM firing procedure, which was promoted as an alternative to the force-defense constituent of the *Safeguard* program.[14] The latter proposal entails, in a much accentuated form, the dangers that are in fact inherent in the doctrine itself, dangers that derive from its unidimensional nature. These and other limitations of the Assured Destruction doctrine, only one of which has in any way been remedied by the post-1969 variant enunciated by the Nixon administration as Strategic Sufficiency, are discussed below.

Assured Destruction and Its Limitations

It is a matter of record that, under the aegis of the Assured Destruction doctrine, expenditures imputable to strategic-nuclear forces have been stabilized; that significant, if not necessarily balanced, arms control measures have been agreed on with the Soviet Union; and that no nuclear war has taken place. It is equally evident, however, that the doctrine, and the force postures derived from it, have serious limitations. Deterrence in itself has certain well-known limitations; exclusive reliance on deterrence in a total strategic system implies additional risks; owing to its implied counter-societal targeting, the Assured Destruction doctrine has its own specific limitations; and current strategic deployments as constrained by the SALT I accords have additional limitations that are not inherent in either deterrence or exclusive reliance on it, or even by the doctrine in itself.

Inherent Limitations of Deterrence

Theorists and defense planners are fond of pointing out that deterrence is a "psychological thing," but in practice, immersed as they are in the technical, bureaucratic, and political problems of the trade, they tend to neglect the psychological core of deterrence, whose problems are so intractable, and in any case hardly amenable to quantitative expression. The complex interactions of rival groups of men upon which successful deterrence must depend may be summarized as follows:

- The deploying party's prior perception of a threat;
- The development and continuous maintenance of reliable strike-back forces capable of inflicting damage deemed to be unacceptable by any and all potential attackers;
- The potential attackers' recognition of the causal link between the promised retaliation and the specific hostile actions that the deploying party is seeking to avert;
- The concurrence of *all* potential attackers with both the technical and the relative-value judgments made by the deploying party on their behalf.

All potential attackers must believe in the virtual certainty of retaliation both in technical terms and as a matter of political decision, and they must also agree that such retaliation would destroy greater values than those which they could achieve by making the move that the deploying party wants to deter.

It is apparent that these (necessary and sufficient) conditions of successful deterrence are more easily fulfilled in the context of an intracultural than intercultural conflict. Without elaborating further, it may be pointed out that a noncultural "rationality" constraint (in the formal sense of means/ends alignment) is indispensable to successful deterrence. In practice, of course, a Western-materialist scale of relative values is now imputed to any potential attacker. If not, he need not be deterred, regardless of the scale of damage that retaliation may inflict. If the tangible values of life and property are less important to a potential attacker than the intangibles of a transcendental faith, a valiant self-image, or a glorious martyrdom, deterrence will fail, as it has so often failed before, in the nonnuclear arena. It may be true that the import of cultural and other value differences will, in each and every case, be nullified by the awesome power of thermonuclear warheads, but it need not be so.

The remote contingency that any one of the many polities whose international conduct is now generally qualified as disruptive or irrational will one day acquire nuclear forces, would be less disturbing if (1) it could be assumed that leaderships of the established nuclear powers would behave "rationally" at all times; and (2) if rationality were enough to ensure the reliability of deterrence. With respect to the former question, what, for example, would be the command and control arrangements operative in a China caught in the throes of another "cultural revolution"? (Or in the midst of an untidy succession crisis?) Or, more speculatively, what of the conduct of a Soviet Union in the process of multinational disintegration?

In any case, as it has been pointed out,[15] deterrence presumes not merely rationality but a given character disposition on the part of the rival leaderships. In order to harness the irrational[16] deterrent threat to a (perfectly rational) policy of deterrent suasion, it must be presumed that the potential attacker will be sober and deliberate (and hence deterred) while the victim will be entirely reckless in unleashing his retaliation. This reverses the normal character attributes of victims and aggressors.

Since survival is at stake, the fragility of the psychological basis of deterrence should not be accepted with the equanimity that has now become customary. On the other hand, it is apparent that the only total substitute for deterrence would be a total defense, and at present such defense is not feasible, nor, in view of the startup risks, desirable. But principal reliance on deterrence in the central strategic relationship with the Soviet Union does not in principle exclude either the deployment of offensive forces other than those needed for simple deterrence or the deployment of "thin" (though comprehensive) population defenses directed at other actual or potential nuclear powers. Nor does it necesarily require rapid-reaction forces permanently aimed at adversary cities.

Additional Limitations of an Exclusive Reliance on Deterrence

In principle, any total strategic system must ensure national security in a narrow, virtually physical sense, and it should provide an additional ("disposable") element of power applicable to political purposes, other than self-defense narrowly defined. Except insofar as particular interests may be equated to national survival, by virtue of their recognized symbolic nature (e.g., the status of Berlin), it is clear that reciprocal (or "mutual") deterrence alone can only provide for self-defense. In fact, it has become the orthodox view among Assured Destruction theorists that strategic nuclear power (presumably short of a disarming first-strike capability) *cannot* be converted into positive leverage in international politics. This view has been controverted in detail elsewhere,[17] where it is argued that even as between the super-powers, strategic-nuclear power has in fact been applied to political purposes and can still be so applied.

Recognition of the value of offensive forces for purposes other than the simple deterrence of direct attack forms the basis of the doctrinal variant of Strategic Sufficiency; this, and the wider issue of the political (and military) applicability of strategic-nuclear power, are discussed elsewhere (see below).

Reliance on deterrence in the central strategic relationship with the Soviet Union, as an alternative to sole reliance on defense, is not incompatible with the unilateral or bilateral deployment of "thin" or selective defenses, against ballistic missiles or other systems. In order to accept exclusive reliance on deterrence, it must be assumed that only "rational-materialistic" powers will acquire (long-range) nuclear weapons, and that they will behave consistently at all times. (Or, alternatively, that in each and every case the R&D or acquisition warning-time with respect to others' weapons will exceed the lead-times of development and deployment for ballistic and/or air defenses). There is, however, a more salient danger than "rational" use: that of a technical error, leading to the launch of any one of the many thousands of thermonuclear warheads now targeted continuously upon adversary populations.

Since the issue of accidental firings has been discussed in great detail in the past, and since an authoritative attempt to revive the debate has been made recently,[18] only one thing need be said on this score: the possibility of an accidental attack cannot be excluded a priori. The confident rejection of the possibility of error is therefore unwarranted. And yet, in view of the possible destruction effects of even a single errant weapon, this possibility must indeed be excluded in order to justify sole reliance on deterrence.

Aside from the contingencies of "irrational" use and unauthorized firings, the self-imposed denial of "thin" (BMD) population defenses, and the deemphasis on aerodynamic defenses, also have a bearing on the processes of proliferation. Specifically, the lack of any defense implied by exclusive reliance on deterrence sets a low entry price on a deterrent relationship with one or both superpowers. There are genuine "potentials" which do have a security motive to deploy nuclear weapons against the Soviet Union. In addition, other "potentials" may be motivated by diplomatic (status) considerations to acquire nuclear arsenals.

Both groups of "potentials" would be effectively discouraged from deployment if superpower population defenses set a high (but not too high) entry price. At present this is not so. Moreover, both superpowers are vulnerable to threats that deterrence alone may not suffice to avert. In this respect, the survivability criteria that are thought essential in the case of the forces deployed by the established nuclear powers may not delimit the status nor negate the threat that small "first-strike-only" offensive forces could present. Their vulnerability would indeed render them "useless, provocative and dangerous," to use the epithets once addressed to the *force de frappe*, but the use of these adjectives would not suffice to contend with, say, a Libya that had somehow acquired control of a handful of missiles.

A small and vulnerable force of known location *could* be destroyed by a police-action attack, but even if effective preemption could be totally assured, which is not likely, the political and psychological costs of such a use of nuclear weapons would be awesome. As against this, even the thinnest of "thin" defenses could provide high levels of security against a force developed, purchased, or suborned by a (small) "disruptive" power. By the same token, it would also reduce the incentive to proliferation, when it is one of the superpowers which would be the intended target.

The Particular Limitations of Assured Destruction

The central objective of Assured Destruction as a doctrine of exclusive deterrence is to reduce, and if possible nullify, the incentive to a (disarming) first-strike attack by the Soviet Union; this is to be done by minimizing the *difference* between the potential destruction inflicted by American offensive forces before and after such an attack. In practice, this was to be achieved by the deployment of powerful retaliatory forces secured against a surprise attack by means of protective construction, mobility, and concealment, rapid-reaction firing, or any combination of these. Further, partly in order to avoid the ("destabilizing") targeting of the adversary retaliatory forces themselves, the doctrine emphasized the targeting of cities, the objective being to destroy the adversary's population, or at least to render his society "unviable." This doctrine is therefore based on exclusive deterrence in its most extreme mode: the value to be threatened is survival itself, and no defense is permissible since it would only attenuate the reciprocal terror that is considered essential to ensure stability. The limitations of the latter aspect of the doctrine have been reviewed (see above), but neither deterrence per se, nor exclusive reliance upon it, requires that the retaliatory threat be so extreme. That is particular to the doctrine.

There is no need to elaborate on the moral implications of counter-societal targeting, nor to reiterate that it can only be irrational in the formal sense of the word. There is also, however, a political problem: the doctrine is addressed to the maximal threat of an all-out attack upon the United States; in theory, therefore, the system is quite inflexible since it could leave a president "with only one strategic course of action, particularly that of ordering the mass destruction of enemy civilians and facilities . . . [or capitulation]."[19] Fortunately, the doctrine has never been implemented consistently;[20] as a result, the offensive forces deployed under its aegis have not been quite as flexible as the preamble to the Strategic Sufficiency doctrine seems to imply. It is relatively simple, if not inexpensive,[21] to modify the target program-ming of (land-based) missile forces so as to allow for small-scale attacks, and for attacks against "military" targets, especially if large, "soft," and fixed. It is apparently to these objectives that the doctrinal variant of Strategic Sufficiency is addressed. More generally, the force deployment guidelines derivable from it would endow the offensive targeting system with additional capabilities to be applied to both political and military purposes. But to do this, a "surplus" in the way of offensive power is required, since any notional application of strategic power would have to leave in reserve all that is deemed necessary for direct deterrence. In practice, the availability of any such disposable offensive power depends on the overall counterforce capability of the offensive forces deployed by the Soviet Union, which is substantial and increasing under the provisions of the SALT I accords. As a result, the doctrinal variant of Strategic Sufficiency is not consistent with the realities of the current strategic balance.

Even with the additional flexibility that Strategic Sufficiency calls for, the prime target of American strategic power would remain the people and society of the Soviet Union. And insofar as the Soviets have accepted the tenets of Assured Destruction, as some now claim,[22] the system would still rely on the reciprocal promise of mass destruc-tion. In principle, this need not be so, but it is obvious that a *unilateral* shift to the targeting of "military" objectives (other than retaliatory forces) could entail unacceptable political costs. It may be argued that in a crisis a United States whose weapons were targeted (and known to be targeted) on military camps, naval bases, airfields, etc., would be in a weaker bargaining position than a Soviet Union whose weapons were aimed at American cities. On the other hand, in view of the continuing SALT relationship, strategic deployment policies need no longer be framed unilaterally, nor should the retaliatory system be static, given that targeting changes can be made rapidly. The alternatives to counter-city targeting are discussed below; here it suffices to say that there seems to be no compelling reason for the permanent and continuous targeting of cities, especially in view of the always present possibility of unauthorized firings.

Additional Limitations of the Current Force Postures (as Constrained by the SALT I Accords)

In principle, the Assured Destruction doctrine does not prescribe force defenses, against ballistic missiles or otherwise. On the contrary, since the purpose of force defenses can only be to increase the reliability of retaliation, their deployment is perfectly consistent with the doctrine. Nevertheless, the SALT I accords (in line with American desiderata by all accounts) limit not only the doctrinally proscribed population defenses but also force-defense BMDs. It is known that in internal debates, force defenses have been opposed by Assured Destruction theorists on the doctrinal grounds that they may entail an "illegitimate" population-defense element, unintended or otherwise. It must be presumed that this was the principal rationale for the SALT I limitation on force-defense BMDs. But the fear that force-defense BMDs could be abused to provide illegitimate population defenses seems inconsistent with the high degree of "transparency" otherwise implied and required by the SALT I accords.

Taken together with the high ceilings set on offensive forces, and with the lack of qualitative limitations, the effects of the force-defense BMD provisions of SALT I are (1) to provide a further incentive to warhead subdivision, and (2) to accentuate the value of rapid-reaction firing. This is so since both remain permissible ways of enhancing the strike-back potential of offensive forces in the face of the substantial, and growing, counterforce capabilities implied by the allowable SALT I force-levels.

In this as in other respects, there has been an erosion of the principal virtue of the doctrine, its emphasis on strike-back capabilities which imply the possibility of deliberation, (international) communication, and (national) control before retaliation is authorized. The dangers of this erosion were illustrated vividly in 1969 when prominent, if unofficial, Assured Destruction theorists allowed themselves to be associated with proposals for a "launch-on-warning" firing doctrine which was promoted as a substitute for a force-defense BMD.[23]

Aside from the limitations of the current force posture (and SALT I) which were discussed above, there are two further drawbacks that do not derive from deterrence, exclusive reliance on deterrence, or the doctrine per se. The first constitutes a unilateral, national problem: the visible imbalance in the force-levels allowable to the two sides. It was of course characteristic of the essentially apolitical perspectives of many Arms Control theorists that they should have accepted the numerical imbalances codified in the SALT I accords with equanimity. As argued in detail elsewhere,[24] their position ignores the profound political significance of third-party perceptions of these gross numbers. Politicians do not measure strategic power in terms of technical force-effectiveness indices with the aid of slide rules.

The strategic balance undoubtedly remains one of the principal variables of international politics, and because of SALT I, political leaders the world over now know that while the Soviet Union is allowed a total of 1,618 ICBMs, the United States may deploy only 1,054; and while the Soviet Union can deploy up to 740 SLBM tubes, the United States is allowed only 656—and conversion options pro rata. None of the sophisticated force-effectiveness criteria put forward by pro-SALT analysts, valid though they may be in purely technical terms, can compete with the simple and compelling message of Soviet superiority that these numbers spell out. Moreover (owing largely to administration statements made before Congress), the world has also been made aware of the fact that some Soviet missiles are distinctly more "powerful" than any deployed by the United States. As a result, from Bonn to Tokyo by way of New Delhi political leaders now speak of the SS-9 as a token of impending Soviet strategic superiority.

As against these vivid impressions, which inevitably modify the political perceptions which in turn guide the conduct of policy, complicated calculations that stress a compensatory warhead, megaton equivalent (ME), or accuracy advantages for the United States in SALT I have no real impact. Moreover, even if relative throw-weight comparisons are not studied in any detail by third-party political leaders, the latter are generally aware of the fact that the warhead potential of the Soviet Union can only be greater than that of the United States. And while bankers discount future monies, politicians anticipate future power trends. The effect of such an anticipation can only be to induce third-country leaders to conciliate the Soviet Union, naturally at the expense of the United States. In this respect, it should be noted that manned bombers, a class of armament in which the United States retains a wide margin of superiority, are widely perceived as obsolete by third-country observers.[25] Deterrence may be a "psychological thing" but it is certainly a "political thing" too. As such, political perspectives, and not those of the systems analysts, should guide force-deployment policies and the conduct of Arms Control negotiations.

The second drawback of the SALT I accords may at first sight appear incongruent with the first: the very high *absolute* level of the allowable offensive forces, as well as the wide scope left open for further expansion within the few parameters constrained by the accords. On the one hand, it is apparent that very low figures, even if attainable given the dynamics of a continuing competition, would also have been undesirable: any depreciation of the strategic-nuclear attributes of superpower status would tend to encourage proliferation on the part of such "potentials" as may have diplomatic/global motives for acquiring nuclear weapons. In a world where the deployment of, say, 100 ICBMs, SLBMs, and bombers sufficed to give a new entrant parity status with the Soviet Union and the United States, the net incentives to proliferation would be high, since the entry price would be low. At present, on the other hand, the very high plateau of the superpower balance has the effect of suppressing the (strategic) status ambitions of others. In this respect, the disappointing diplomatic returns of the British Independent Deterrent and of the *force de frappe,* generally attributed to their small size, have had a cautionary effect on these particular potentials—though not, of course, on those whose motives for acquiring nuclear weapons derive from considerations of local or regional security.

But in view of the very high allowable offensive force-levels of SALT I, and of the so far unimpended scope given to accuracy improvement and warhead subdivision (which is, of course, encouraged by the limit placed on the silo envelope parameters), the danger of reducing the price of superpower status is now very remote. Instead, there is the economically undesirable and, more important, strategically dangerous, multiplication of warheads, if not launchers. It is evident that any new doctrine, and any new strategic relationship conducted under its aegis, must attempt to facilitate and encourage limitations in the scale of the offensive forces deployed on both sides under conditions of *visible* parity.

The Constituents of a New Strategy

In its concentration on the avoidance of deliberate and central nuclear war, the Assured Destruction strategy reflects the priorities of its time. In the later fifties and sixties the principal strategic danger seemed to be the possibility of a "mechanical" breakdown in the balance of terror, caused by the destabilizing interaction of the rival deployments. At this time also, because of the general political immobilism of a frozen bipolarity, the strategic-nuclear competition frequently occupied the center of the stage in superpower relations. From the Missile Gap deception of the late 1950s to the Cuba Missile Crisis and beyond, the highest officials of the land were often preoccupied by fairly narrow technical issues concerning the deployment or movement of a few dozen missiles on the part of one side or the other. Given such high-level attention, what were essentially technical issues naturally became matters of central political concern.

In this setting, the attainment of a disarming counterforce capability could be represented as a reasonable policy goal, or at least one which could be freely imputed to the other side. And the prospect of a nuclear war precipitated by the acquisition of a disarming capability on one side, or the desire to anticipate such acquisition on the other, seemed not implausible. It was to this danger, of an "apolitical" war precipitated by the mechanical imperatives of preemption, that the Assured Destruction doctrine was primarily addressed. The *collateral* danger of unauthorized (including accidental) firings was, in effect, "assumed out" of the system, which implicitly assumed technical and command-mangerial perfection. And more remote dangers, such as those associated with the emergence of new nuclear arsenals in unreliable hands, were totally discounted.

In fact, as argued above, the Assured Destruction doctrine, and the deployment policies formulated under its aegis, did not merely ignore these collateral and secondary dangers but actually accepted their intensification in order to reinforce the stability of the balance of terror. Moreover, the doctrine also implied a total and symmetrical denial of the political utility of strategic-nuclear power, on the assumption that "as between the superpowers" strategic nuclear forces could be used for no purposes other than simple deterrence.

Although declaratory policies and force-deployment decisions did not consistently conform to the doctrine, there is no doubt that the synthetic symmetry of the Mutual Assured Destruction model has eroded the policymakers' propensity to see strategic-nuclear strength as a politically useful form of power which could be applied to purposes other than self-defense, narrowly defined. But the symmetry of the doctrine is not congruent with the military imbalance in Europe. So long as Western Europe remains insecure vis-à-vis the Soviet Union, a disposable surplus of American strategic-nuclear power will remain an indispensable ingredient, in one form or another, of European *political* stability. It is therefore apparent that any reformulation of American nuclear strategy must contend with the political inadequacy of the present doctrine and of the present posture, as well as with the neglected collateral and "secondary" dangers of exclusive deterrence which concern the mutual security of both the United States and the Soviet Union.

There appear to be three major alternatives to the present strategy. The first, a "pure" strategy based on the deployment of comprehensive and "thick" population defenses, has been explicated and refined by Donald G. Brennan over many years.[26] This strategy calls for a total reorientation from deterrence (which is to be totally abandoned) to active defense. This, it has been argued, need not result in a destabilizing Soviet reaction so long as the transformation is accomplished by mutual agreement in the context of a continuing arms control dialogue. (In the past, Dr. Brennan has stressed the strong support given to strategies of active defense by Soviet analysts and policymakers.) Aside from the superior reliability of an active defense which does not rely on the fragile and delicate interactions of deterrence, it has been claimed that a defensive strategy would also facilitate and promote effective measures of arms reduction since "thick" defenses could absorb the asymmetrics that would inevitably occur in the discrete stages of an arms reduction process, and since they would negate any "cheating."

A second alternative strategy accepts the premises of Assured Destruction but would reduce substantially the scale of the offensive forces deployed. This, a strategy of Minimum (or "finite") Deterrence, is based on the assumption that much smaller damage levels than those now considered necessary would in fact suffice to deter.

The third alternative is not any single "pure" strategy but a whole class of strategies based on various combinations of limited defenses (both "thin" population defenses and "thick" *force*-defenses) with optional elements such as changes in targeting and firing doctrines, and the adoption of a limited strategic war posture. One such *composite* strategy is presented below.

Mutual Assured Survival and Security (MASS)

It was recognized above that deterrence is, on balance, preferable to a full ("thick") defense in the central strategic relationship with the Soviet Union, where the two are mutually exclusive. But reliance on deterrence does not imply in itself either the *continuous* counterpopulation targeting or rapid-reaction firing procedures which enhance (1) the possibility and (2) the destructive impact of unauthorized firings.

Counter-Population Targeting. When intercontinental ballistic missiles were first deployed their median miss distances were of the order of several miles; only "soft" and/or extensive objectives could provide feasible targets for their warheads. Even with median miss distances of five miles, unprotected ICBM sites could in fact be targeted, but it was the larger adversary cities that made the natural targets of early ICBMs. In any case, less extensive targets, such as army camps, submarine bases, naval yards, and transport infrastructures, were not feasible targets, or at least they were not efficient targets since they did not maximize the destructive impact of the available warheads.

Today, however, ICBM median miss radii are no longer measured in miles: the figures commonly quoted for the newer (U.S.) long-range missile systems are of the order of 1,500 feet or less, and it appears that further significant reductions are in prospect even without major innovations.

In *political* terms, the rationale for countercity targeting has always been that in a "deep" crisis, where the interests in dispute are deemed "vital," the threat of a countercity strike would be the most effective bargaining weapon. In such a context, typical of the maximalist scenarios that once animated strategic thinking, it is obvious that if the weapons of the Soviet Union were aimed at adversary cities (and could not be retargeted) while those of the United States were only aimed at, say, army bases, the latter would be at a serious disadvantage. But in terms of confrontations which are not quite as "deep," where the interests in dispute are less than vital if still very important, it is apparent that the threat of a "no-cities" strike would be more effective: since the damage threatened would not be quite as catastrophic, the threat would be more credible.

In *strategic* terms, the doctrine emphasized counterpopulation targeting because its crucial mechanism was reciprocal terror; the more intense the terror, the more powerful the mechanism, and no terror could be more intense than that induced by the threat of (an urban) genocide. Since countercity targeting was in some sense at the opposite extreme from the doctrinally proscribed targeting of the offensive forces themselves, the whole thrust of the doctrine promoted the former, and gave it an almost virtuous quality. It is apparent, however, that there is no logical connection between the avoidance of counterforce deployments and the targeting of adversary cities if warheads are available for both.

In *technical* terms, the requirement of early ICBM and SLBM systems for fairly rigid predetermined targeting programs ensured that, once cities were targeted for the reasons given above, they had to be targeted *continuously,* since targeting changes could not be made very rapidly and reliably.

But the present political, strategic, and technical environment has largely invalidated these rationales for continuous countercity targeting. In political terms, the current phase of detente, and the related decline in the centrality of the strategic-nuclear dimension of international politics, has sharply reduced the incentives to countercity targeting. Detente is compatible with renewed confrontations over less-than-vital interests but it is certainly not congruent with the kind of "deep" crisis in which the threat of a countercity strike remains even residually credible. The sort of crisis which may still precipitate a direct confrontation between the superpowers, over, say, the control of Persian Gulf oil, could still conceivably provide some scope for limited strategic nuclear attacks, but not for credible massive threats against cities. An American declaratory threat to, say, Soviet naval bases in Iraq, or even in the Black Sea, may retain some measure of credibility, but not a threat to Moscow or Leningrad.

In *political* terms, therefore, the present environment renders countercity targeting ineffectual in the face of all contingencies except for a parallel threat to American cities—and it is very difficult to imagine circumstances in which such a threat could reflect any rational purpose of Soviet policy. In *strategic* terms, countercity targeting does not effectively absorb offensive forces away from counterforce deployments. At present levels of warhead deployment, and even without the further warhead subdivision that is likely to occur, even a modest portion of the superpower offensive arsenals would suffice to destroy all worthwhile population targets. Moreover, in *technical* terms, the availability of reliable, semiautomatic, and virtually real-time retargeting systems means that continuous advance targeting is no longer required—even if cities were to remain the ultimate targets.

At present a "no-cities" targeting doctrine for U.S. strategic offensive forces would therefore be: (1) technically feasible (with facilities for retargeting against cities if and when required), and (2) strategically neutral with respect to the counterforce issue. It is argued here that a "no-cities" targeting pattern would actually be desirable on political grounds, if the present detente continues.

If cities are not targeted, and if counterforce targeting is rejected as technically too exacting and politically counterproductive, two alternative *unilateral and declaratory* targeting policies come to mind: (1) "deep sea" (no preset targeting at all) and (2) "military" targeting. Under (1) the United States would, in the first instance unilaterally, declare that its offensive forces would no longer be targeted against adversary cities, and that the latter would not be targeted again in the future unless "crisis conditions" seemed to be imminent. At the same time, it would be made clear that survivable facilities for reliable retargeting against cities would be maintained. Aside from the not necessarily negligible psychic and propaganda gains of such a declaratory move (which might well prompt a parallel Soviet declaration), it could also serve as an intermediate step for the second option. The "military" targeting option (2) properly pertains to the wider issue of the political application of strategic-nuclear strength, in the context of the (growing) disequilibrium in (nonnuclear) military strength between East and West with respect to Europe. (For a discussion of this option see below.)

It is apparent that no verifiable bilateral agreement on a "no-cities" targeting pattern is feasible with present means of external verification. And it is also apparent that any internal means of verification would be altogether too intrusive to be acceptable to either side, and certainly to the Soviet Union. Hence a "no-cities" (and no strategic counterforce) targeting posture would have to be evaluated in purely unilateral terms, with no reciprocity expected. This is a sector in which a "soft" nonverifiable agreement would not be desirable, if only for political reasons.[27]

With respect to the danger of unauthorized firings, even a unilateral "no-cities" posture would have the advantage of materially reducing the destructive impact of an errant weapon, and hence the possibility that it would trigger a (miscalculated) Soviet response. Needless to say, it would be altogether more comforting to know that the continuous targeting of American cities was also suspended, and thus an errant Soviet weapon would not impact over a populated area.[28]

Rapid-Reaction Firing. If the strike-back capabilities of the present strategic offensive forces were compatible with slow, and therefore deliberate firing, procedures, the balance of terror would be much less delicate than it is, and the danger of unauthorized firings correspondingly smaller.

But in spite of all the expenditure that has been devoted to enhancing weapon survivability over the last two decades, the bomber element of the offense remains critically dependent on ground alert and a rapid takeoff (followed, however, by positive controls on weapon release). As for the land-based missile element, this could become dependent on rapid launch between attack waves (subject to "pin-down") since Soviet progress with warhead subdivision techniques may force the abandonment of the present postattack firing doctrine. Even the least time-dependent element of the triad, the submarine-borne missile force, could be required to fire rapidly in certain circumstances, i.e., if SSBNs were being lost to an ongoing Soviet strategic ASW offensive.

While protective construction, dispersion and alert, concealment and mobility are the key variables in the survivability of land-based ICBMs, bombers, and SSBNs, respectively, rapid-reaction firing procedures remain a valuable additive factor in each case. And they would so remain regardless of any additional protective measures. Only a drastic reduction in the counterforce potential of the rival offensive forces could reduce, and even eliminate, the incentives to rapid-reaction firing. Subject to unexpectedly successful progress in SALT II (or perhaps SALT III) the independent progress of guidance technologies and the related increase in the scope for subdivision seem certain to increase the counterforce potential of U.S. offensive forces. That of the Soviet offense has of course a much greater scope for expansion, given the very considerable throw-weight advantage accorded to the Soviet Union under the agreements now in force. This trend is bound to increase the incentive to rapid-reaction firing procedures still further, particularly on the American side (given the throw-weight imbalance).

There is, however, an alternative to a drastic balanced reduction in the throw-weights of the rival offensive forces that now seems improbable in the medium term—even if the SALT negotiations continue successfully. This would be to separate the "legitimate" weapons from those deployed against the strategic offensive weapons themselves. At the point of launch, it is of course impossible to make any such distinction, but at the point of impact the distinction is very clear: a warhead that impacts on an ICBM site or bomber base would certainly be a counterforce weapon; a warhead that impacts on a city is not. Point defenses that can absorb attacks against strategic offensive forces but not attacks against other targets, could make the distinction effective, and thus provide a substitute for that drastic reduction in the counterforce potential of both sides that cannot now be achieved by bilateral agreement.

Effective force defenses, deployed on an agreed bilateral basis and in the context of the SALT relationship, to ensure "objective" perceptions of their precise scope, purpose, and limits could thus reduce and hopefully eliminate the incentives to rapid-reaction firing. Moreover, point defenses for fixed-site ICBMs could also reduce the incentives to the deployment of mobile missile systems. While mobility would certainly enhance weapon survivability against pretargeted ballistic attack, it is generally recognized that the deployment of mobile ICBMs would also prejudice the prospects for arms control in the medium term.

Force defenses have of course been evaluated in the past against alternative (passive) defensive measures. While the intricate question of the relative cost-effectiveness of alternative active and passive force defenses cannot be discussed here, it would appear that the principal "arms control" objection to active force defenses is no longer valid. The contention that force defenses would be perceived as retaliation-negating population defenses, and would therefore provoke additional offensive deployments, is inconsistent with the nature of the present strategic dialogue with the Soviet Union. "National" means of verification would suffice to discriminate quite unambiguously between (static) site defense and area defense interceptors and radars, while the SALT dialogue would enable both sides to communicate precisely the scope and limits of any deployments.

With respect to the land-based missile element, effective force defenses could virtually nullify the incentive to rapid-reaction firing procedures; they would also reduce the incentive to the maintenance of a high alert status for any sheltered bomber force. (Additional measures, such as controls on the deployment of attack submarines, would be needed to reduce the importance of rapid-reaction firing procedures in the case of SSBN forces.) It is apparent, however, that it is the land-based missile element that would benefit most, and this would be a valuable gain in itself since it would enhance the relative worth of this class of weapons, and correspondingly retard the deployment of mobile systems, including additional submarine systems. Since *no* class of weapon is more amenable to "external" inspection, it is apparent that retarding the obsolescence of the land-based missile element would facilitate progress towards further measures of arms limitation and reduction. And "technology control" is of course the key to successful arms control in the long run.

A second element in the new strategy should therefore be the deployment of force defenses in order to absorb counterforce capabilities and thus reduce the incentives to rapid-reaction firing as well as retard the obsolescence of fixed-site and hardened ICBM forces. In any case, the present, virtually "hair-trigger" posture, with the resultant maximization of the possibility of unauthorized firings, cannot be accepted with equanimity as a proper and normal component of the strategic balance in an era of detente.

Selective Deterrence. When it became apparent that China would acquire nuclear weapons and, eventually, long-range delivery systems of reasonable efficacy, a debate began between the advocates of exclusive reliance upon deterrence and those of a second school of thought which accepted the imperatives of deterrence vis-à-vis the Soviet Union but argued that the inclusion of China in the deterrent arena was neither inevitable nor desirable. When the *Sentinel* BMD system was unveiled in 1967 and promoted as an "anti-Chinese" population defense, the second school of thought appeared to have prevailed. But subsequent developments indicated that the "Chinese" rationale was more of a public relations stance than an expression of stable strategic thinking.[29] In any case *Sentinel* included a force-defense element too. The "Chinese" rationale was promoted in the political context of frozen U.S.-Chinese hostility, a hostility that has now given way to the most promising of all the current detentes.

Having registered the legitimate doubts that may be entertained about exclusive reliance on deterrence in general, it must be recognized that if the Soviet Union is deemed an acceptable partner and antagonist in a deterrent relationship, China ought to be so accepted too. While Chinese capabilities in the command and control sector are likely to be less reliable than those of the Soviet Union, this technical factor is outweighed by political considerations: something of a U.S.-Chinese alignment is plainly emerging and China is likely to add much more to its anti-Soviet deployments before initiating significant strategic-nuclear deployments against the United States. As a result, the Chinese strategic nuclear effort is eventually likely to result in a significant detraction from Soviet nuclear power, and this is a detraction that the United States can only view with favor.

Although both the Soviet Union and China are vulnerable to major internal upheavals which could have incalculable side-effects in the strategic-nuclear arena, both are established and historic polities whose coherence and continuity in foreign policy—dictated by geopolitical externalities—have always exceeded their fragile internal stability. This coherence is not to be found in the case of many of the new states, including some that could acquire nuclear weapons in the medium or longer term, whether by domestic development, purchase,[30] or subornation. The doctine of Assured Destruction would have the United States accept a deterrent relationship with any and every possible nuclear power, it being tacitly assumed that only powers provided with survivable strike-back forces would qualify. This qualification, the only one made, and typically apolitical, assumes that first-strike-only deployments would simply be preempted, at any rate if the fingers on their trigger were deemed to be "irrational." This, however, would require a most difficult political decision, and the theory ignores the very heavy domestic and external political costs of preemption. Such costs may well outweigh the strategic imperatives to preemption, especially since the decision to strike will inevitably seem to be an open option, not time-limited (until the target acquires strike-back capabilities), while the costs of an attack would of course be immediate.

Since even a crash program with limited ambitions would take many years to develop in the case of the "possible-potentials" that can now be envisaged, the danger that intercontinental weapons will fall into unreliable hands remains remote. On the other hand, the development and deployment of even the "thinnest" of comprehensive population defenses (and the required prior renegotiation of the ABM Treaty) would also take the best part of a decade. Hence a doctrinal change and a reformulation of policy would have to be made in the near term. This is so especially since if it were known in advance that minor-country nuclear forces would be effectively negated, this could have a *preventive* effect on third-country acquisition policies. In particular, the incentives to acquisition could be sharply reduced in the case of potentials with extra-regional ambitions if it were kown that any would-be "global" nuclear deployment would be negated in advance by superpower population defenses. The latter intention would also imply a much greater likelihood of eventual preemption on the part of either or both superpowers, thus reducing the incentives to acquisition still further.

Needless to say, the Japanese nuclear weapon program, the current national forces in Europe, and any furture joint-European forces would *not* fall in the category of the forces to be negated. Assuming that present political alignments persist, any Japanese or European weapons would be deployed against the Soviet Union (and China); these detractions from the disposable "net" of Soviet strategic power, even if minor, would enhance the relative strategic position of the United States. By the same token, such third-country forces would of course degrade the relative position of the Soviet Union, but this is a loss that the Soviet Union would be bound to accept. Both Japan and any national European forces would of course satisfy in full the (minimal) stability qualifications required for admission into the deterrent arena.

The doctrinal change here advocated is therefore intended to make deterrence selective, and specifically to restrict exclusive reliance upon it to relationships with the major power centers, and these alone. In practice, this would imply the declared willingness to deploy comprehensive (albeit "thin") population defenses intended to nullify small-scale strategic-nuclear threats emanating from any new nuclear powers whose development potential is limited. While it may be comforting to believe that nuclear power breeds responsibility, the physical survival—and political flexibility—of the United States should not have to depend on any such unsubstantiated belief.

The BMD controversy of 1969-70, and the prior, largely internal, debates of the 1960s, saw a large number of arguments deployed for and against particular BMD systems and the BMD concept in general. Without reviewing these debates, it appears that the principal strategic objection to the population-defense BMD concept was, again, the fear that any population-defense would provoke a destabilizing Soviet response, or even invite preemption. There were also technical objections, based on the purported inefficacy of defenses in general and ballistic missile defenses in particular. It is apparent that once the finite scope of population defenses, and of their antiballistic missile element, is limited to minor-power threats, such technical objections would become invalid (though not of course the budgetary objections). As for the dangers of a destabilizing Soviet reaction, it has already been argued above that the SALT relationship would preclude any real misperceptions of the ultimate scope of "thin" population defenses. In any case, no such defense is here advocated unless its deployment is preceded by a negotiated agreement with the Soviet Union, which would no doubt imply reciprocity.

The Political Desiderata of MASS

If from the bilateral viewpoint of mutual survival and security both superpowers can be said to have too much in the way of offensive power, from a unilateral political point of view the United States may also be said to have too little. There is no doubt that the very large throw-weight capacity allowable to the Soviet Union under the terms of the SALT I accords affords much greater scope for warhead subdivision than is the case with the American offensive arsenal. In strategic terms, this implies the possibility of a serious threat to U.S. strike-back capabilities—but only if, and when, this potential is actualized and subject to any corrective restraints that may be agreed upon in SALT II, III, and so on.

But in political terms the situation is entirely different. Virtually ignoring the warhead count criterion (and for that matter the "megaton-equivalent" criterion too) observers in Europe as elsewhere have perceived the SALT I accord in political terms, as is appropriate. The unequal SALT I force-levels, stated in terms of simple (and therefore compelling) launcher nose-counts, have spelled out a message of American inferiority, a message reinforced by the particular qualities of the Soviet arsenal, i.e., the large size of some of the allowable launchers. Political observers do not rely on slide rule computations of relative force effectiveness; nor do they take balancing factors outside the scope of the agreement into account (especially since the bomber element of the balance is normally discounted on the technically invalid but psychologically persuasive grounds that "bombers are obsolete"). This is at any rate the impression conveyed by official European statements. (See, for example, the 1972 German Defense *White Paper*.)

It may seem that once again "strategic" questions that ought to be left to the strategists are being misconstrued by unqualified political observers. But *strategy is politics,* and even if technically invalid, third-party perceptions are fully valid in political terms: at bottom, the willingness of the United States to concede the apparent superiorities of the Soviet Union inscribed in the SALT I accords did reflect domestic political pressures for disengagement and for a decrease in the level of military preparedness. Accordingly, SALT I was in fact an act of retreat even if (1) the superiorities accorded to the Soviet Union are more than outweighed by relative American advantages in other dimensions of strategic power, and (2) it were indeed true that such superiorities are "meaningless" in strategic terms.

Since the perceptions of political leaders abroad determine their conduct, even technically invalid misperceptions of SALT I translate into tangible political losses for the United States. Such losses are apparent in the further impetus that has been given to the disintegration of the Western "Alliance" (now to be placed firmly within quotation marks). It may be that these losses are outweighed by gains in the U.S.-Soviet detente—which SALT I may be said to have facilitated—and it is also possible that even *net* political losses are outweighed by "strategic" or even budgetary gains within the strategic-nuclear arena itself. It is obviously pointless to attempt a comparitive evaluation of such gains and losses.

In any event, the SALT I accords and the resultant balance of *perceived* strategic nuclear power represent an irreversible reality in the short term. It is moreover exceedingly doubtful whether SALT II (or III) could correct the launcher throw-weight imbalance in order to bring about a visible parity between the superpowers. Nor can the American advantage in warhead numbers be fully developed as a (medium-term) visible counterweight to ostensible Soviet superiorities, since this would conflict with the imperatives of a MIRV limitation agreement. Without the latter, assuming that the Soviets do deploy *some* fairly accurate multiple warhead system successfully, the American land-based missile force would be rendered unacceptably vulnerable.

But even within the SALT I limitations and all the other constraints (i.e., congressional) that are now operative on American deployment policies, it may still be possible to enhance the political utility of American strategic-nuclear power. In particular, it appears that the basic asymmetry in the overall military posture of the superpowers allows scope for a reapplication of American strategic power to political purposes, beyond self-defense narrowly defined. This is because the Soviet Union remains the preponderant power in Europe, while the United States is not similarly placed with respect to any areas that are vital to the security of the Soviet Union.

In the strategic-nuclear realm it has long been recognized that effective leverage depends as much on credibility as on effective capabilities. A technically formidable but less credible threat may count for less in the life-or-death nuclear arena than a threat which though technically modest is also highly credible. There is therefore a balance of "resolve" as well as a balance of technical power, and resolve is not a psychological quality which can be simulated at will, as some would have had us believe, but rather a reflection of perceived interests. Though diplomatic manipulations may certainly heighten their perceived centrality, it is the nature of the *interests* at play that will determine perceptions of resolve, and thus delimit credibility.

In this respect, it is apparent that even for a somewhat disengaged United States, the security and independence of Western Europe represents a central interest. This is not so for any major Soviet interest now vulnerable to American (nonnuclear) action. In the strategic-nuclear arena, this gives the United States an important *organic* advantage that may even outweigh a considerable degree of strategic inferiority. Unfortunately, in the case of European defense, the first use of American strategic nuclear power under the present strategy is decreasingly credible. This, and the decline in the relative strategic capabilities of the United States exemplified by the SALT I accords, has played a major part in inducing West Germany and other NATO members to conciliate the Soviet Union—at the expense of joint "alliance" interests.

The lack of credibility that now limits the "convertibility" of American strategic nuclear power in the European security context is not, however, an inevitable consequence of the present superpower balance of forces. After all, the European nuclear guarantee retained a substantial measure of credbility until quite recently, long after it was no longer associated with any specific element in American strategic capabilities.

The recent decline in the credibility of the U.S. nuclear guarantee has derived from the reasonable political assumption that the United States would not in fact expose its population to Soviet attack by launching an attack of its own upon the population of the Soviet Union (in retaliation for an invasion of Europe). To a lesser extent, it also reflects the reasonable *technical* assumption that the United States does not have sufficient "disposable" power, and in particular the means required for a preemptive counterforce strike—even if partial—against Soviet strategic forces.

It is in this context that the "no-cities" targeting proposal discussed above should be seen. If in terms of *declared* policy the United States were to establish a targeting continuum (in respect of European defense) between the presently deployed tactical weapons and the strategic offensive forces—against an ascending hierarchy of (nonstrategic) "military" targets—this would admittedly erode the tactical strategic "firebreak" that some still consider residually useful, but it could enhance very considerably the political utility of American strategic power.

The declaratory policy which is here advocated would link the use of American strategic weapons against *relevant* (nonstrategic) Soviet "military" targets to particular acts of aggression against European NATO members. By the same token, the intention to use American strategic-nuclear weapons against Soviet cities would be disclaimed, except, of course, in the event of a Soviet attack on American cities. In this scheme of things, it would be understood that the Soviet Union would have full reciprocity, at least in theory: in the event of an American (or NATO-American, but not European-only) attack upon Eastern Europe, the Soviet Union would be free to use its strategic weapons against *relevant* American "military" targets (or irrelevant military targets if it so chose). Any use of Soviet weapons against American *cities,* on the other hand, would of course result in a MAD-style retaliation, for which a reliable capability would be retained.

Given the basic asymmetry between the Soviet Union, which is still residually revisionist, and the West, which is committed to no more than a European status quo if that, it is apparent that the United States could gain much greater benefits from such a "no-cities" declared targeting policy than the Soviet Union. And the Soviet Union could not counteract this advantage except by attempting to claim the right to use its (nonstrategic) military power against Western Europe, while threatening retaliation against the American population for any use of strategic weapons against nonstrategic Soviet "military" targets. Such a claim would have to be *declared,* and in the present political context it seems most unlikely that it would be asserted. The detente which the Soviet Union is so assiduously promoting is perfectly compatible with the defensive application of American strategic power to Europe in a "no-cities" targeting framework, but it cannot be compatible with the assertion of a Soviet military prerogative in Europe that denies the American strategic option while keeping nonnuclear (and tactical nuclear) attack options open to the Soviet Union.

It is recognized that aside from the implied erosion of the tactical/strategic "firebreak," there are other weighty objections to this proposed strategy. Domestic (including congressional) objections are certain to materialize, though it is by no means clear that such objections would be either very intense or very widespread, given that the targeting of Soviet "military" objectives would operate in a strictly defensive framework, and that the entire strategy would be associated with a "no-cities" targeting pattern.

In Europe itself, it may well be that the wider problem of reestablishing American prestige in its nonmilitary dimensions would be the overriding factor in validating the new strategy in political terms. And of course it is European perceptions and the European response that would determine the success or failure of the new strategy. It is also possible that aside from the natural and understandable resistance to the new policy that must be expected from the Soviet Union, there could also be a disruption of the whole process of detente *within* the Soviet Union.

In addition to these macro-level objections, there are also some narrower but highly significant technical problems including the difficulties of discriminating between "military" and city targets from the *victim's* point of view, the only one which is relevant. And there is also the lesser problem of differentiating between "military" targets and strategic-force targets. But the attempt to validate strategic power in political terms does not require full compliance with every technical desideratum. But it does require that strategic policy be made in a *political* context, a context that stresses continuous political effects rather than the increasingly remote and ever less credible contingency of nuclear war.

Notes

1. At a later stage, of course, continental air defenses acquired an additional role in protection of SAC bases. In 1953 the RAND Corporation circulated a preliminary study of the issue of bomber/base vulnerability, R-244-S. In 1954 the full version of the study, Report R-266, "Selection and Use of Strategic Air Bases" (RAND, Santa Monica) was circulated.

2. Secretary Dulles's speech is reproduced in U.S. Department of State *Bulletin,* Vol. 30 (25 January 1954), pp. 107-110.
3. For a review of the debate, see R. E. Osgood and R. W. Tucker, *Force, Order and Justice* (Baltimore, Johns Hopkins University Press, 1967), pp. 121-192.
4. Reality remained more complex; during the McNamara years even the staunchest advocates of the doctrine did not attempt to achieve consistency by scrapping the air defenses or relegating their role exclusively to the protection of the bomber bases. (Force-defenses, unlike population defenses, are of course consistent with the Assured Destruction doctrine.)
5. In the summer of 1961, the United States had deployed a total of 63 ICBMs, as against fewer than 50 for the USSR; and 600 B-52s, 1,000 B-47s, and 30 B-58s as against a Russian bomber force estimated at 70 Tu-20s, 120 Mya-4s, and 1,000 Tu-16s. *The Military Balance,* pp. 2, 3, 8, 9 (London: ISS, 1961). In subsequent years, as the emphasis shifted in the offense sector of ICBMs, the *margin* of American superiority increased further, reaching a peak, in gross vehicle numbers in 1966. The United States then deployed a total of 2,126 units and the USSR, 625. If qualitative factors were taken into account, the margin would be still greater. *The Military Balance,* p. 67 (London: ISS, 1972-73).
6. S. P. Huntington, *The Common Defense* (New York: Columbia University Press, 1961), pp. 47-64.
7. Contrast the carefully qualified position taken by Secretary McNamara on the political utility of strategic power with the second quotation below. "A . . . substantial numerical superiority of weapons [sic] does not effectively translate into political control, or diplomatic leverage. . . . Today, our nuclear superiority does not deter *all* forms of Soviet support of communist insurgency in Southeast Asia . . . we, and our allies as well, require substantial nonnuclear forces in order to cope with levels of aggression that massive strategic forces do not in fact deter." (Emphasis added.) From the San Francisco speech. See his remarks before UPI editors and publishers (18 September 1967) reproduced *inter alia* in "Scope, Magnitude and Implications of the United States Antiballistic Missile Program," *Hearings before the Subcommittee on Military Applications of the Joint Committee on Atomic Energy 90th Congress,* First Session, p. 108. "Strategic nuclear weapons, certainly as between the superpowers, are good for nothing, but mutual deterrence." McGeorge Bundy at the R.I.I.A., London, 1972. See the text of a lecture reproduced in *The World Today,* Vol 2, No. 28 (February 1972):52.
8. The initial inability of Secretary McNamara to appreciate the *political* significance of the Russian missile force that was being deployed in Cuba ("what difference does it make if the missiles are launched from Cuba instead of the Soviet Union" conveys the tenor of his stance) is a good indicator of McNamara's tendency to discount the political aspect of things strategic.
9. I.e., a Counterforce virtually stripped of its political purpose in inter-Alliance relations, since it no longer bolstered European defense. Russian strategic offensive force were still the targets, but primarily in the context of an ongoing Russian nuclear attack upon the United States, and in order to reduce its destructive impact.

10. For all the wealth of numerical data, the level of damage deemed to be "unacceptable" was inevitably quite arbitrary. Any level of damage could be represented as "unacceptable" so long as consistent assumptions were made with respect to the scale of relative values imputed to Russian leaders. Prudent men could be represented as being effectively deterred from launching an attack—in pursuit of whatever (nonsurvival) interest—by a reliable retaliatory threat to a single city. A martyr/irrational leadership, on the other hand, an analogue to Hitler raving in his bunker, would demonstrably remain undeterred by a reliable threat to 99 percent of the Russian population.

11. The latter was contingent on the scale of the adversary counterforce capability and that of any strategic defenses. In theory these were "hard" variables (derived from intelligence projections and the resultant computations of ICBM kill-capabilities and attrition rates in penetration). In practice, these variables proved to be very fluid. Advocates of the doctrine opposed the force-defense element of *Safeguard* even in the presence of convincing data on the prospective vulnerability of the U.S. ICBM force.

12. Because the damage levels thought necessary with respect to the Soviet Union were much higher than those applicable in the Chinese case.

13. As pointed out above, the scale of the damage to be inflicted varies with the relative-value scale imputed to the adversary leadership.

14. For the proposals, and those who made them, see P. W. Wolfowitz, "The Proposal to Launch on Warning," *Hearings: Military Procurement for Fiscal Year 1971* (U.S. Senate Committee on Armed Services 1970), pp. 2278-81.

15. See F. C. Iklé, *Every War Must End* (New York: Columbia University Press, 1971), pp. 118-131, for a discussion of the issue of deterrence on which my own remarks are based.

16. In the formal sense: killing x percent of the Russian (or Chinese) population in the wake of a nuclear attack upon the United States would not in any way improve the material position of the nation. On the other hand, the *ability* to inflict retaliation could be used to extract reparations from an aggressor. Hence the act of retaliation—if it follows directly the attack that provokes it—can only be irrational, since means are not aligned with goals.

17. E. N. Luttwak, *The Strategic Balance* (Center for Strategic and International Studies, Washington, D.C.: Georgetown University Press in association with the Library Press, 1972), pp. 69-96.

18. F. C. Iklé, *Can Nuclear Deterrence Last Out the Century?* California Arms Control and Foreign Policy Seminar (January 1973).

19. *U.S. Foreign Policy for the Seventies* (III). A Report to Congress by Richard M. Nixon, President of the United States (9 February 1972), p. 157.

20. The duration of deployment programs has generally exceeded the lifespan of the strategic doctrines enunciated by successive administrations. As a result, the official doctrine, and the force postures maintained under their normal aegis, have usually been somewhat out of phase.

21. The "Command Data Buffer" now being developed by the USAF for its ICBM force has a total planned acquisition cost of $286 million. *Department of Defense Appropriations for 1972,* Hearings Before a Subcommit-

tee of the House Committee on Appropriations, 92nd Congress, 1st Session, Part 5 (1971), p. 1139. In any case, the force already has a relatively flexible reprogramming facility.

22. See, for example, the statement of the principal (official) U.S. SALT negotiator, Ambassador Smith, before the Senate Armed Services Committee in which he asserted that Russian acceptance of the ABM Treaty signified their acceptance of the doctrine. *Military Implications of the Treaty on the Limitations of Anti-Ballistic Missile Systems and the Interim Agreement on the Limitation of Offensive Arms,* Hearings Before the Senate Armed Services Committee, 92nd Congress, 2nd Session (1972), pp. 383-384. It may be, however, that Russian acceptance of the doctrine is in fact temporary and contingent on the quality of the BMD technology available to the Soviet Union. Certainly, the ideological fervor that animates the proponents of Assured Destruction in the United States is hardly likely to be duplicated in the Soviet Union, where even more hallowed doctrines are said to be on the wane.

23. Luttwak, *The Strategic Balance.*

24. Ibid., pp. 69-96.

25. Whether *new* aircraft, with new names, would be perceived as valid weapons is an open question.

26. See, for example, D. G. Brennan, "The Case for Population Defenses," in *Why ABM?* ed. J. J. Holst and W. Schneider, Jr. (New York: Pergamon Press, 1969).

27. Because of the impact on Western opinion, where it would be liable to undermine realistic attitudes towards the necessity of preparedness. There would be no counterpart pressures in the Soviet Union.

28. In the absence of any data (in the public domain), the reliability of Soviet command and control systems remains an open question.

29. At one time, in 1967, it seemed possible that the lack of any BMD program could prove politically embarrassing.

30. Not the purchase of complete forces or even weapon systems. As the abortive Egyptian missile program showed, it is quite easy to recruit mercenary scientists and technicians and to acquire many weapon components from commercial sources. That both the hardware and the skills so acquired were inadequate in the Egyptian case was due to funding limitations more than to political obstacles to access. Assuming that oil-generated funds were to be made available in amounts ten or twenty times larger than those invested by Egypt in the sixties, it seems probable that both warheads and missiles could eventually be produced, albeit not very sophisticated ones. But then ICBMs need not be very sophisticated to be effective even against the United States, if the latter is entirely undefended against missile attack. Israel may well attempt to preempt any Arab nuclear weapon program and either super power could do so too, but this would not necessarily nullify the danger. The undoubted technical ability to monitor such a program, and if need be, preempt it, does not automatically imply the political ability, or will, to do so.

2

The Problems of Extending Deterrence
(1980)

Extended Deterrence in American Politics

It has been reported that on the eve of his assumption of office President Carter asked the Joint Chiefs of Staff to supply him with an estimate of the minimum number of strategic nuclear weapons that would suffice for deterrence; the Chiefs were specifically invited to consider the possibility that 200 ICBMs might be enough.

Even without knowing the exact form of words employed by the president, it is a fair speculation that the deterrence he had in mind could only have been simple, strike-back-only deterrence—that is, deterrence for the self-protection of the United States *strictu sensu*. It will be recalled that during that same period President Carter was insistent on expressing his devotion to the North Atlantic Alliance; in fact, he was then actively sponsoring an all-for-NATO defense policy whose immediate implications were the (abortive) decisions on the withdrawal of American troops from Korea and the redeployment of major naval units from the Pacific to the Atlantic. There was much emphasis on the need to strengthen NATO's (conventional) military strength.

Subsequent events show that it would be wrong to attribute sinister motives to the conjunction of presidential enthusiasms; no conscious intent to decouple European security from strategic nuclear deterrence is to be imputed. One is not dealing with holy selfishness, nor indeed with strategic logic of any kind. It is rather a case of unstrategical pragmatism, apt to pass virtually unobserved in a political culture itself profoundly unstrategical. For the president NATO is one subject and strategic nuclear policy quite another; why confuse matters by connecting the two? When strategic nuclear questions are considered it is the techniques and tactics of the mechanical interactions of the two

35

forces, and above all the possibilities of arms control, that dominate attention in a framework naturally bilateral. Equally, when NATO and its travails are considered the business-like thing to do is to focus on the administrative detail of "rationalization."

President Carter has been allowed to remain perfectly consistent in his inconsistency: he does not appear to have declared himself even once on extended deterrence, under whatever name. Certainly, he has not emulated his predecessors, who used to reiterate with some frequency the nexus between the security of the Alliance and the *ultimissima ratio* of strategic nuclear punishment.

The context, so far, of the SALT II debates in the U.S. Senate shows that President Carter's inadvertence on the question of extended deterrence cannot be counted as one of his eccentricities. A great deal has been heard of the Allies during the July 1979 hearings before the Senate Committees on Foreign Relations and the Armed Services but, with the single exception of Henry Kissinger, no witness and no senator has expressed any reasoned position on the role of American strategic nuclear weapons in the deterrence of Soviet military action against NATO Europe. Certainly, there was no depiction of the part that strategic weapons might play in inhibiting Soviet attacks against European-based nuclear weapons—this being, of course, the most obvious intraconflict function of extended deterrence.

Instead, there has been a great deal of discussion of Allied perceptions of the strategic balance and of the impact of ratification—or its denial—upon those perceptions. As used in the debates, the word *perceptions* is invoked to explain why things that *do not* make a difference in "reality" are nevertheless important. Thus Secretary Brown, in explaining why he felt it necessary to react to the Soviet acquisition of a partial counterforce capability (against the American *Minuteman* force), first denied that such a capability made a "real" difference, and he then went on to argue the need to match it for the sake of the beneficial effect on third-party "perceptions."

What makes everything so complicated is the highly specialized meaning of "reality," itself defined in terms of the theory of Mutual Assured Destruction. According to that theory, which remains devastatingly influential even when ostensibly repudiated, *no* meaningful advantages can be derived from the acquisition of strategic nuclear capabilities in excess of the mathematically prescribed requirements of assured destruction and below the (very much higher) requirements of a fully disarming counterforce capability. True believers still firmly deny the need for any reaction to the projected emergence of sundry Soviet superiorities in the ICBM sector. Secretary Brown and his colleagues also explicitly reject the contention that the Soviet Union could extract any "real" advantage from the acquisition of a *partial* counterforce capability (against the American *Minuteman* force). Specifically, they have ridiculed various crisis/blackmail scenarios put forward by Paul Nitze and others. But they part company from the true believers by nevertheless advocating the need for the MX program, explicitly to achieve perceptual effects which they hold to be necessary to maintain essential equivalence.

It was only in this psycho-political realm that the connection between the central balance and the Alliance was recognized during the July 1979 debates. The Allies were accorded the role of important spectators whose opinions of the central balance would be very important, but there was no suggestion that the central balance might actually affect their security directly by conditioning Soviet conduct—no suggestion, that is, that the spectacle might actually have a concrete effect on its audience.

Thus the question of extended deterrence has not so far figured in the debates on the SALT Treaty. "Perceptions" always excepted, minimalists and maximalists agreed to treat the central balance as a closed bilateral system. The pro-Treaty minimalists have denied that discrete shifts in the strategic balance in favor of the Soviet Union would have any significance whatever, since the United States would still retain an Assured Destruction capability; the anti-Treaty maximalists argued that these adverse shifts were highly significant, since they revealed that the Soviet Union was attempting to attain a fully disarming counterforce capability. What seemed to be at issue was the reliability of self-protecting deterrence (as well as perceptions of the same) and not the scope of deterrence.

For the minimalist the omission of extended deterrence from the argument is a matter of strict logic: the theory of Mutual Assured Destruction inherently defines a purely bilateral world in which the values of allies, and even the protection of United States forces overseas, is only possible to the extent that these things are fully assimilated into the body of national territorial interests (so that attacks upon them would warrant an automatic retaliatory response). For the maximalists, it would be equally logical to define *desiderata* in terms of the requirements of extended deterrence, but in fact they have failed to do this. With rare exceptions, they define the significance of increments in Soviet strategic nuclear capabilities strictly within the closed bilateral framework, saying little or nothing about third parties.

Since it is only at the level of strategical thought that the connections between things are revealed, it is not surprising that extended deterrence should have been the ignored issue of the SALT II debates, even though it could have served very well as an evaluative criterion of genuine significance.

The Theory

It is against this sobering background of political inadvertence that the theory of extended deterrence should be reviewed.

The mechanism whereby intercontinental nuclear deterrence is extended to offer protection to Allies and American forces overseas (and thus beyond the minima set by national boundaries) is, of course, the mechanism of escalation. It addresses the total *potential* scope of weakness at conflict levels below the intercontinental nuclear—that is the sum total of values whose protection cannot be adequately provided by either defense or deterrence at each of those lower levels. It is obvious enough that protection obtained from extended deterrence must always be a second-best solution, for there is always some inherent reluctance to escalate and this automatically offers a corresponding opportunity for coercive diplomacy by the other side. What counts, of course, is the degree of this reluctance, which naturally varies sharply from time to time and more gradually over time with changes in the balance of intercontinental nuclear vulnerabilities. More precisely, the *actual* scope of extended deterrence is defined by the interaction of two quite different balances: the balance of relative intercontinental nuclear vulnerabilities on the one hand, and the "balance of perceived interests" on the other. It is the interaction of these two balances, as viewed from both sides, that determines the credibility of escalation in any one particular case and thus determines the scope of extended deterrence.

The connection between the degree of absolute vulnerability of the would-be protector and the scope of extended deterrence *ceteris paribus* is obvious enough: the more vulnerable the would-be protector, the less convincing is his promise to mete out punishment, given that retaliation is expected. But that is only half the story, since the risk of extending deterrence also depends on the expectations of the other side: the more vulnerable the antagonist to be deterred, the less likely he is to invite punishment in pursuit of a given objective. For each side there is the further consideration that, in facing a conflict which is apt to continue, the amount of damage that can be inflicted on the other side is a significant value in itself—on the assumption that the damage expected at this stage is still very far from the entirely catastrophic. It is for these reasons that the balance of *relative* vulnerabilities is the operational criterion on the risk side of the equation.

The factors that govern the balance of relative vulnerabilities are physical in nature and easy enough to define, if often very hard to measure. These factors are: the capabilities of each side's forces (offensive and defensive); the quality of the civil defense and industrial recovery organization and the scope and robustness of passive defences. Certainly, it is not the simple balance of intercontinental nuclear forces that defines the risk side of the equation.

By contrast, the factors that govern the "balance of perceived interests" are political and diplomatic, atmospheric and therefore evanescent. The intensity of each side's interest in a given value can change drastically over time (for example, the American interest in maintaining West Berlin's independence from Soviet control increased very sharply between 1945 and 1961), and, of course, there is always much room for error in each side's assessment of the intensity of the other's interest in a given extraterritorial value. Thus the balance of perceived interests, of sentiments and their reciprocal assessment on each side, is a complex and uncertain thing, sentiments being, of course, variable over time, while assessments may lead, lag or be plainly in error.

Indeed, error is unavoidable. The intensity of a given interest may increase sharply and precisely in response to a challenge—thus retroactively turning what was a correct assessment into an underestimate. Conversely, the unmet challenge stimulates retroactive justification through the downgrading of the interest. The process that determines the relative value of a given thing to its would-be protector is, of course, political in the broadest sense, with the dynamics of internal politics often being activated by external challenges. Thus the evolution of the Berlin value, which was finally equated with a national-territorial interest (President Kennedy's "I am a Berliner"), owed much to the internal political reaction to external challenges, French as well as Soviet. Equally, nonterritorial and nonsubstantive values can be devalued by oblivion—that is to say, by tranquillity.

In addition to the declaratory dimensions of commitment making or disengagement, there is the cumulative effect of diplomatic action and reaction. Urgent and hard responses to attempted encroachments or implied challenges will strengthen the true intensity of the interest protected, as well as affecting assessments of the same. Hesitant and yielding responses will, by contrast, affect the balance of perceived interests and adversely, increasing the risks of extended deterrence much as a deterioration in the balance of vulnerabilities might. Diplomats may be forgiven if they attach greater value to the benefits of flexibility, while, for the most part, strategists are more mindful of the added risk of misperceptions that a flexible stance entails.

Extended Deterrence Today

The Case of Europe

As far as Central Europe is concerned, extended deterrence now comes into play in two very different circumstances: in negating the effects that the Soviet ability to attack European cities with nuclear weapons would otherwise evoke, and in deterring a Soviet disarming offensive against European-based theatre nuclear weapons. Given that European-based nuclear weapons do not threaten Soviet cities as convincingly as Soviet medium-range ballistic missiles (MRBMs) and intermediate-range ballistic missiles (IRBMs) and bombers threaten European cities, the latter must be hostages unless protected by the extended deterrence of American intercontinental nuclear weapons. It must be assumed that the British and French independent deterrent forces only protect British and French cities if they protect anything at all. The rationality assumptions of deterrence itself define the problem as a matter of denying opportunities for coercive diplomacy rather than one of actual protection. Does American extended deterrence still meet the need? Will it continue to do so given an American-Soviet balance of vulnerabilities that increasingly favors the latter? The answer must be yes, although for reasons not altogether reassuring.

In this case it is the balance of perceived interests that dominates the matrix. On the one hand, there is the enormity of the deed and the lack of Soviet gain in actually executing it; on the other, there are the sentiments that would be evoked. More realistically, the latent threat to the cities is dominated by the availability to the Soviet Union of other (and infinitely more plausible) instruments of military pressure at much lower levels of conflict intensity. American extended deterrence thus meets only a very feeble requirement in this case: to negate a blackmail option, which must rank very low in the Soviet repertoire. Thus even the advent of an increasingly dubious parity should not abridge the effective scope of extended deterrence in this case. This should apply equally during conflict: only a Soviet Union in the process of being defeated in combat at lower levels of intensity might plausibly threaten European cities to force a cessation of hostilities. In the above case it is implicitly accepted that the Soviet threat to European cities would suffice to inhibit a NATO attempt to impose a cessation of hostilities (at lower levels of intensity) by threatening to attack Soviet cities (for example, with the *Poseidon/Trident* warheads allocated to Supreme Allied Command, Europe). It is not, of course, the recent sharp increase in the relative vulnerability of the United States that would determine the result, but rather the longstanding, absolute vulnerability of the European cities, including those that the British and French nuclear forces might protect by deterrence in other circumstances.

In this case, therefore, we find extended deterrence to be as fully applicable, as credible and as stable as it has ever been; but this happy situation reflects weakness rather than strength—specifically the wide range of other opportunities for Soviet coercive diplomacy in peacetime and the utter implausibility of a victorious NATO offensive (or for that matter counter-offensive) in wartime.

The Case of Japan

In the case of extending deterrence to negate Soviet nuclear threats to the population of Japan, matters are not quite the same. Here too the sheer enormity of the deed must greatly inhibit Soviet action, quite independently of any fear of direct retaliation. But, of course, there is still a difference between the prospects of the victim and those of the aggressor, and here too the difference must be equalized by extended deterrence to deny a corresponding potential for coercive diplomacy. A significant difference between the two cases is the fact that the Soviet Union does not have, vis-à-vis Japan, a comparable repertoire of high-intensity threats (while having an even broader repertoire at the low-intensity end of the spectrum). Specifically, the Soviet Union's ability to carry out a successful amphibious invasion of Japan is altogether more dubious than her ability to advance against NATO forces in a nonnuclear conflict. This would increase to a corresponding extent the saliency of nuclear threats in a crisis of sufficient severity.

It follows that the requirements which American extended deterrence must meet are somewhat more demanding in this case. Since the Soviet Union is acquiring such abundant intercontinental nuclear forces, the fact that American extended deterrence must cope with a threat now emanating from Soviet weapons of lesser range does not alter the character of the problem. Nothing much would be gained by deploying American weapons designed to be directly comparable with the SS-20 and *Backfire*. What *would* make a great deal of difference would be an American counterforce advantage, for this would provide a plausible and straightforward instrument of deterrence in this case: silo-killing ICBM warheads would be much more plausible as instruments of extended deterrence in this particular case than less accurate dedicated weapons of lesser range. (In nature, range and accuracy go together, but given the deployment choices actually available, the two would not be congruent in this case.)

Deterrence by Counterforce

The wider question of the role of counterforce capabilities in extended deterrence is best discussed in the more specialized context of the Soviet threat against European-based theatre nuclear weapons.

The Soviet ability to attack those weapons has greatly increased of late, and it is still increasing. The difference between the new SS-20 and the old SS-4s and SS-5s is not merely greatly increased accuracy but also greatly diminished vulnerability. Because of their vulnerability, SS-4s and SS-5s could scarcely be used in a selective, discrete fashion: all would be immediate candidates for destruction if any were used. SS-20s, by contrast, are mobile and easily concealed, so that selective attacks by them are a true option. Even though its present warheads are relatively large, the SS-20 has a potential for high accuracies that would allow the useful deployment of low-yield warheads suitable for preemptive precision attacks upon Quick Reaction Alter (QRA) air bases, *Pershing* predispersal depots and any future ground launched cruise missile (GLCM) basing points (the currently envisaged USAF deployment scheme does little to minimize the threat).

Given the role that these theatre nuclear weapons play in the structure of NATO security, even a fully equivalent countervailing threat (against Soviet MRBMS, IRBMS, and G- and H-class submarines and nuclear-armed aircraft) would not dispose of the problem. In the NATO structure conventional forces serve to deter incursions and to defend against conventional attacks; battlefield nuclear weapons serve to deter a full-scale conventional offensive and to defend against one that appears to be within measurable distance of success, as well as to deter the prior use of Soviet battlefield nuclear weapons. Theatre nuclear weapons in turn serve to deter Soviet attacks against NATO battlefield nuclear weapons, as well as to defend against a successful Soviet offensive that features the use of nuclear weapons of any kind.

Unless they are fully confident of NATO's self-deterrence, Soviet war planners must undertake to peel away these layers of protection one by one: superiority in nonnuclear offensive potential is their base capability, and some sort of equivalence in battlefield nuclear weapons is sufficient for the Soviet Union, given the former advantage. But as long as American extended deterrence covers European cities, the deterrent effect of European-based theatre nuclear weapons cannot be negated by Soviet threats of retaliation against European cities. Nor can the Soviet Union achieve her goal by threatening reciprocal use, for Soviet theatre nuclear attacks upon NATO forces in the process of being defeated would have by no means the same significance as NATO theatre nuclear weapon attacks upon victorious Soviet forces. It follows that the logical Soviet goal must be to acquire a disarming counterforce capability against European-based theatre nuclear weapons.

In theory, this particular threat could be negated by a combination of active and passive defences in addition to dispersal and mobility, the major means of protection for these weapons. In practice, however, the *unalerted* theatre nuclear-weapon force is highly vulnerable. Of particular concern is the fact that the refinement of Soviet military capabilities provides the possibility of a *nonnuclear* disarming offensive, in addition to the already manifest trend towards a precise nuclear offensive offering the prospect of low collateral damage.

The latter threat already generates a requirement for extended deterrence quite distinct from the much less demanding requirement associated with the threat of nuclear attacks upon European cities. Very specifically, it is a selective counterforce option that is called for. In this case the straightforward balance of American and Soviet offensive forces is of central significance to the scope of extended deterrence. The emerging Soviet advantage in ICBM counterforce capabilities is undoubtedly depriving the United States of the most effective instrument of deterrence vis-à-vis the threat of a disarming counterforce attack upon the European-based theatre nuclear forces. The reciprocal threat of selective strikes against Soviet ICBMs loses plausibility when the Soviet Union can count on achieving a net improvement in the residual balance by a (nonescalatory) response of identical form. Given an American advantage in ICBM counterforce capabilities, the deterrence of low-casualty Soviet attacks against European-based theatre nuclear weapons could be accomplished by the threat of low-casualty American attacks against Soviet ICBMs or theatre nuclear assets. Given the deterioration in the ICBM balance, the scope of extended deterrence has effectively been abridged.

Nor would the deployment of the 2,000-warhead MX force which is now envisaged suffice to restore the situation. This would undoubtedly reestablish the ICBM component of the "triad" for the purposes of self-protecting deterrence, but it could not restore the American ascendancy in the ICBM sector of the competition, given the planned characteristics of warheads and guidance. Nor can compensation be found in other categories or weapons. In the case of air-launched cruise missiles, the selective use of the force is improbable on operational grounds alone; it is also apt to be counterproductive, given the vulnerability of the bomber bases, including dispersal bases, to which the B-52 is confined. As for submarine-launched ballistic missiles (SLBMS), neither the *Poseidon* nor the *Trident* has the requisite counterforce characteristics, while the inherent counterforce potential of the large-diameter tubes in the *Ohio*-class submarines remains as yet unexploited.

What applies to the extended deterrence of Soviet attacks upon theatre nuclear weapons based in Europe is true, a fortiori, of the extended deterrence of such threats as a Soviet invasion of northern Norway or a seizure of Berlin or attacks upon lesser allies overseas. In all such cases the threat of selective counterforce attacks would in the past have provided the most plausible deterrent, one that would capitalize on the very great political and pyschological difference between threats to uphold and threats to change the status quo. In both cases, there is not now a conventional defence nor a battlefield nuclear deterrent of sufficient plausibility. In both cases only the hope that the Soviet Union might anticipate wider consequences that she would judge to be broadly unfavorable stands between these exposed positions and Soviet power.

As the overall military balance continues to change adversely, it is important to remember that the outpost can receive deterrent protection from the prospect of a wider conflict only to the extent that the *initiation* of that wider conflict is more frightening to the potential aggressor than to the victim. In the case of northern Norway there is, in theory, much scope for dealing with the problem *in situ* by endowing the outpost itself with formidable defences and a battlefield nuclear deterrent also.

In the case of Berlin (as, for that matter, in the case of an American aircraft carrier moving within range of Soviet attack submarines) the protection of the lesser thing must derive from the strength of the security system as a whole. Given weakness in the whole, the outpost is correspondingly less well protected from afar. How plausible is it to a potential aggressor that the center would indeed choose to widen a conflict if the expected outcome were no better for the whole than for the outpost on its own? (And, of course, infinitely more catastrophic.)

Here, then, a partial answer may be given to the prescribed questions: if extended deterrence remains as reliable and credible as before in the presence of a deteriorating military balance, it is the stability of the system that is being compromised. In the past military decline would exact its price pro rata in a corresponding loss of control over territory. In modern conditions boundary lines do not adjust as easily to absorb change, for there is no frequent conflict in detail at the margins of civilized states. The delusion that adverse changes in the military balance will have *no* consequences is easily inspired by such conditions. But military decline is as costly as it ever was. Instead of a visible retreat, the result is progressive fragility in the system of deterrence, and this is a process which may long continue to be hidden from view, especially in conditions of political tranquillity, which are themselves contrived for the sake of their strategic consequences.

All remains the same on the surface, while the substance of deterrent protection is increasingly eroded. Eventually, the plausibility of the threats on which deterrence is based may collapse overnight. The exposed positions that are protected only by perceptual lags will then have to be conceded by a prudent victim when the hollow deterrent is challenged. Even then, the results of military decline could still be denied, but only to the extent that the victim can persuasively promise to react to the challenge in a manner reckless and by then self-destructive. Such a policy, whether adopted in protection of Berlin or designed to nullify Soviet counterforce advantages by a "launch on warning" firing doctrine, certainly offers a most economical response to the broad and serious military effort of the Soviet Union—but then the stability of the entire system would be sacrificed for the most marginal of our peaceful comforts.

3

The Nuclear Alternatives

Uwe Nerlich's analysis of theater nuclear forces in Europe, a work that is scholarly and of urgent practical moment, and his dismal account of unstrategical conduct on the far side of the Atlantic and of political incoherence in the United States, leads to an incisive assessment of the inadequacies of the status quo. Given the Nerlich definitions of the dissatisfactions that prompt the quest for alternatives, one can proceed straightforwardly to examine what those alternatives may be. Such an examination, however, must be prepared for by a brief analysis of some aspects of the West's strategic circumstances.

In a tranquility that may in part be contrived, the West is drifting toward a situation in which the residual agency of its political confidence vis-à-vis the Soviet Union may be nothing more substantial than the perceptual lags that blind it to its own weakness. As for physical security, the West would have to count to a large degree on the continued prudence of a future Soviet leadership of which nothing is known.

It is worth recallng that the powerful reluctance of Soviet (and indeed Russian) rulers to *initiate* conflict derived from circumstances and not from some inherent quality of restraint. In the past, Soviet leaders operated in a climate that combined long-term optimism and operational pessimism: they could believe that their society and economy would progress beyond ours, and at the same time they had little confidence in the ability of their soldiers to execute combat operations with the swift and controlled elegance that is obviously required to extract advantage from war, while avoiding widespread nuclear destruction. Their long-term optimism systematically diminished the incentives to initate any conflict, since at any one time the future power balance could be expected to be better than that of the present, but their operational pessimism was of course the great disincentive.

Nowadays those circumstances no longer apply. As those people who would minimize the gravity of our military inadequacies insis-

47

tently remind us, the Soviet Union faces a bleak future, and it is their argument that we should subtract that future from the actuality of our military disadvantage when we assess the overall balance of power. But perhaps the more correct procedure would be to multiply rather than to subtract. The classic conditions of war are present in the conjunction, though of course NATO may not be the target of the first onslaught. Consider the situation in dynamic terms: the long-term pessimism that a chronically shoddy industry, increasingly unfavorable demographic trends, and an ideology in decadence might justifiably induce is surely a powerful incentive to act before it is too late. Politicians everywhere are reluctant to begin a decisive action that is replete with dangers, no matter what the advantages dangled before them might be. But the one frequently persuasive argument is the choice between action now or never to bring unfavorable competition to an end by the use of force.

At the same time, the disincentive of Soviet operational pessimism must have waned very greatly. In Ethiopia we saw the successful accomplishment of a most elegant military operation, the entire character of which was decidedly un-Russian. Instead of the broad certainty of land contiguity, there was only a thin line of supply by air and by sea, and those supply lines were easily interrupted. General Petrov, presiding over his Cubans and Eastern Europeans, embodied in his own person the advent of a new era of Soviet operational confidence. Granted, the Somali forces are not to be compared with NATO forces or with the Chinese, but consider the setting: Soviet military officers were introduced into a shrinking Amhari enclave, deep in the African interior, to try to save the situation with enemies converging from all directions. Perhaps this new and most dangerous feeling of control and self-confidence may yet be eroded by the Afghans' heroic struggle. But we do not know who initiated the first Afghan adventure or which particular group of Soviet decision makers made the fatal promises of swift success.

I would thus argue that the unfavorable military trends—whose direction and magnitude have notably ceased to be controversial even though their meaning remains so—unfold at a time when the whole character of Soviet decision making may be changing in an adverse direction. It would not be the first time in this century that a conflict was precipitated by leaders who tried to find a solution in war for economic decline, elite cynicism, and government insecurity and who used war to preserve a multinational empire whose demographic trends were adverse and whose subject nations had become decreasingly controllable in peace. It is after all quite rational to act before decline reduces the opportunities for action; power used in the present may yet alter an unfavorable future by altering the terms of the competition.

In the past, when most fighting was more localized and on a smaller scale, military decline exacted a proportionate price in the shrinking of the boundaries of control; nowadays the national boundaries are quite rigid and do not change easily to accommodate shifts in the balance of power. Instead, military decline now results in a hidden erosion of the reliability of defenses and the stability of deterrence, a process that may go on for a long period of time without manifest effect. Even while the inadequacies of our defenses become more definite and even while the gaps in deterrence continue to widen, we continue to seek safety in declaratory policies (which promise a reckless reaction such as "launch on warning") and in the promise of abrupt and self-destructive escalation. By such means we strive to spare ourselves the burden of matching the Soviet Union's broad effort to achieve preponderance in one dimension of military power after another. Without having to make any sacrifices we can thus continue to uphold our political independence, and we may even still feel quite secure despite our inadvertence. But of course it is the stability of deterrence that we are sacrificing, that is to say, the very substance of the system from which our safety derives.

The Waning of "Extended Deterrence"

What, then, is to be done? Briefly recalling the position—we now find ourselves in a situation characterized by the waning of "extended deterrence." Variously influenced by temporary financial stringencies, disjointed reactions to the Indochina conflict, a pervasive failure of strategic intellect, and the misplaced hopes of arms control, successive U.S. administrations have allowed the West's greatly superior intercontinental nuclear capabilities of the later 1960s to wane into a dubious parity and that parity is now to be conceded. It is not necessary to recite the familiar numbers, and it is even less necessary to argue the detailed computations: the matter was fully and starkly presented in the SALT II hearing before the U.S. Senate. As the last residual advantages wane, the United States will soon find itself sharply inferior in both area-destruction and silo-killing capacities.

It is true of course that, even before, American policy failed to uphold an operational doctrine for the use of intercontinental nuclear forces that would specifically exploit the superiorities of the weapons to extend deterrence. It is also true that the United States has long since lost the perfect immunity that would have made extended deterrence a relatively easy affair. But there was a general sense, strongly felt and soundly based, that the net superiorities of the U.S. intercontinental nuclear forces formed the material basis of extended deterrence for two reasons.

First, some ICBMs could be fired in small numbers, selectively, and without prejudice to the rest of the ICBM force since the Soviet Union had only a slight counterforce capability against all the ICBM silo launchers. In the same vein, the small and technically backward Soviet SLBM force of the 1960s did not effectively threaten the SAC bomber bases, thus again allowing a selective use of some bombers without prejudice to the rest of the force. Overall, the relative magnitude of the forces involved allowed an extensive allocation of weapons to the tasks of extended deterrence (for attack against all Soviet targets except cities), while still reserving an ample and properly diversified force for the *ultimissima ratio* of simple, strike-back deterrence in protection of American soil.

Second, even with immunity lost, the United States had a very great advantage in the balance of relative vulnerabilities that, other things being equal, assured it of a dominant position in each phase of any possible escalation. Extended deterrence is of course entirely derived from plausible declarations of a willingness to escalate; its credibility accordingly reflects the assessments of each side as to its *own* vulnerability (and, indirectly, the vulnerability of the other side), as well as the balance of perceived interests. Because of the adverse change in the ratio of offensive forces, the partial but concrete Soviet advantage in "strategic" air defenses, and not least because of the unilateral Soviet civil defense effort, the balance of relative vulnerabilities has moved decidedly against the United States.

This does not mean that there is no longer any effective protection at all to be expected for NATO Europe from extended deterrence. To the extent that U.S. interest in maintaining the status quo is very strongly felt and so acknowledged and that core values are highly involved, extended deterrence remains persuasive, for escalation is still of course physically feasible. This is the case for example in the extended deterrence of Soviet nuclear attacks against European cities; the sheer enormity of the deed and its lack of inherent gain make the threat implausible, and a postattack reaction of reckless fury remains as likely as it ever was.

But if this rather unimportant class of threats remains firmly negated (and quite useless as an eventual instrument of Soviet coercion) the shrinking of extended deterrence removes protection from a whole class of targets that are altogether more plausible: airfields and large military installations in general—especially in NATO areas that are not protected by nuclear weapons—and theater nuclear weapon basing points in particular. It will be recalled that in the "balance of imbalances" that was, a prominent function of extended deterrence was to protect theater nuclear weapons during actual conflict. So protected, those weapons could in turn inhibit the use of Soviet battlefield nuclear weapons—to deter their first use and to deter their use outside the immediate battle area if NATO had already used nuclear weapons on the battlefield itself.

There is, of course, a whole school of theoreticians and publicists whose stock in trade is the endlessly reiterated claim that strategic nuclear weapons are entirely useless except for mutual deterrence. Having never allowed extended deterrence and its requirements to figure in their models, they do not detect its shrinking in the present circumstances. There is no Europe or NATO in their purely bilateral world in which everything in excess of the force-levels needed for "assured destruction" can be dismissed as mere "overkill." Even now, those two slogans define the whole problem for a good many Americans, and those Europeans who applaud SALT II must logically stand with them. Of course, it is rather more difficult on this side of the Atlantic to sustain the illusion that the exclusive bilateralism of the mutual-assured-destruction school of thought resembles reality. In terms of our problem, matters are quite simply defined: with extended deterrence so greatly diminished in scope, something must take its place or else a number of distinct Soviet threats become credible, with diplomatic consequences even if not actually military ones.

The Adverse Nonnuclear Balance

Is it still controversial to say that the nonnuclear defense of NATO is now less viable than it was 10 years ago? Great as the Soviet effort has been to build up its sea power and nuclear weapons of all categories, both ground and air nonnuclear forces have received an ample flow of resources. These resources have enabled the Soviets to add positive air power to air defense, to complete the panoply of armor-mechanized capabilities with SP guns, to broaden chemical warfare capabilities, and, more broadly, to highly develop the ancillary and supporting forces that had been neglected in the past—from mobility engineering and transport to electronic warfare. On our side, we seem to focus attention on such things as attempting to improve the efficiency of weapons acquisition—in the hope perhaps of emulating the high degree of standardization of the French army of 1940, whose equipment was so much more homogeneous than that of the Germans.

As for the advent of the new-technology precision-guided munitions

(PGMs), only those people who persist in confusing technical effectiveness with operational utility can believe in the great importance of those weapons. PGMs in large numbers may indeed achieve an increase in the attrition exacted by the defense, insofar as tactical and environmental circumstances are favorable. This may yield a degree of ex post facto satisfaction, but it will not decide the contest, for the enemy's operational method is to penetrate deeply by using armor-mechanized forces and then to encircle the opposing forces. The only reward for the divisions that do hold their ground by the efficient application of firepower would be disruption and capture as the enemy completes the encirclement and cuts them off.

For Norway and Denmark, the threat is primarily that of a coup de main in three dimensions. This, as in central Europe, is a maneuver threat, but one that is much less robust, being heavily dependent on the achievement of surprise. It is sometimes forgotten that even without political surprise, there may yet be operational surprise, to be achieved perhaps against forces that mobilized long before and are in the process of demobilization.

But the north Norway sector is an impregnable fortress compared to the central front, where the long and thin cordon of NATO forces could scarcely cope with warfare in the modern style, opportunistic reinforcements of successful axes of penetration. Amored forces with broad capabilities are not to be defeated by the episodic attrition that may be exacted by weapons with only narrow applications; nor can the deeply echeloned columns of armor that would sustain the momentum of a Soviet offensive be defeated by a linear defense that would rely on the mere administration of firepower.

Even this dismal assessment does not include any consideration of the disruptive effects of chemical attacks. Some NATO forces are entirely unequipped to deal with the threat, by self-denying ordinance; others are equipped only defensively and would not be able to neutralize the great operational disadvantages that chemical protection would entail. Their troops, masked and robed, would have to fight those who might be neither.

Also within the sphere of land warfare, we find that the NATO air forces offer a combination of an impressive technical advancement and a diminishing operational utility. Our air forces, themselves so very costly, do force the Soviet Union to divert great resources to providing air defenses, and this reciprocal force-building effect is of incontrovertible value. But the Soviet Union does sustain that effort, it still maintains large offensive ground forces, and it still develops its own positive air power while we are left with air forces that are largely preoccupied with only the suppression and evasion of air defenses. The much less capable air forces of the Warsaw Pact may yet achieve much more in war because our own battlefield air defenses are so much weaker.

It was the great lesson of the 1973 Middle East war that although the physical attrition that Soviet-style air defenses can impose may fairly easily be contained, the effort to do so merely increases virtual attrition. Only the battlefield interdiction mission, accomplished by high speed deliveries of area munitions at very low level, remains a promising avenue of air deployment, but only the Royal Air Force and the Luftwaffe seem interested in following that approach. The willful refusal to acknowledge the lessons of the 1973 Middle East war is not of course due to intellectual blindness; as in the similar failure to contend with the real phenomena manifested in the Russo-Japanese war in Manchuria, it is institutional priorities that suppress the evidence. Certainly the types of aircraft and ordnance acquired since 1973 could not have been justified if the lessons of that war had not safely been explained away as exotic epiphenomena not worthy of serious consideration.

These grave operational inadequacies have scarcely been kept secret, and they must therefore undermine deterrence vis-à-vis an adversary who must surely know better. It is however fair to consider the possibility that our operational ineptitudes may yet be redeemed at the political level. Although NATO's technically impressive PGMs may be tactically fragile and operationally of very limited worth, the Warsaw Pact's Poles, East Germans, Hungarians, and Czechs may be uncooperative and even actually disruptive. But of course it is circumstances that make both saint and sinner: it takes only a passive compliance to follow in the wake of a seemingly victorious offensive, thereby adding to its momentum. Certainly NATO cannot depend on the hope that if it does win the battle, the dissatisfaction of the East European troops could then take concrete form.

Once again complex matters achieve a simple resolution. In this range of conflict we also find grave inadequacies, which are only to be remedied by appealing to the higher court of the battlefield and theater nuclear deterrence and defense.

The Options

Having thus defined the problem, our starting point is radically different from that naturally inspired by the procedures and mentality of arms control. Instead of focusing on European-based nuclear weapons themselves, comparing them directly and unstrategically with those of the Soviet Union, their role is defined in terms of the broad needs of deterrence and defense at all levels of conflict. Instead of agonizing over the difficulties of categorization and enumeration, necessary preludes to the drafting of arms control limits, our concern is with the security needs that must be met and the establishment of priorities among them.

For a good many years now, several conceptually different proposals to improve deterrence and defense have been in circulation, but none meets all the requirements on its own, for none is broad enough to provide remedies for both the shrinking of extended deterrence and the nonnuclear inadequacies. The main proposals are discussed below.

Battlefield Nuclear Defense

Continuously refined by the Shreffler-Sandoval Los Alamos group, this proposal in its latest form still calls for the deployment of large numbers of low-yield warheads to be aimed at front-line enemy forces by short-range missiles with highly accurate guidance systems. If this proposal were adopted, NATO nonnuclear forces would amount to little more than a strong frontier guard.

The intellectual foundations of this proposal were established long ago by Bernard Brodie, and it was his simple contention that a war in central Europe would be quite catastrophic even if entirely nonnuclear. A policy of deterrence based on the declared intention of using tactical nuclear weapons as soon as it was manifest that war was clearly intended would thus be credible—there being little to choose from between the two sorts of catastrophe. Hence Brodie, ardently followed by the Shreffler-Sandoval group, saw no merit in the desire to keep a high nuclear threshold. The Los Alamos group stresses the high cost that traditional forces entail, and its studies are quite acute in their diagnosis of the fundamental weaknesses of the present deployment of nonnuclear forces. They especially point out that incremental improvements in a thin linear defense are inevitably costly (in direct proportion) but do not add to our security (nonnuclear) in proportionate degree, since the line would still remain too weak at any one of the narrow sectors chosen by the enemy for penetration.

The Shreffler-Sandoval proposal takes full advantage of the cost-saving potential of various new technologies, and it offers a full remedy, in principle, for the inadequacies of our nonnuclear forces. On the other hand, it does nothing to remedy the shrinking of extended deterrence, and indeed offers no deterrent protection for anything behind the battlefield itself.

This means that the proposal is incomplete, but that does not in itself nullify its merits, which are considerable. In the first place, if warfare is to be waged by straight attrition, as is currently prescribed in U.S. officer training, nuclear attrition is certainly much more efficient. Second, if missiles are to be used to deliver nuclear warheads against the enemy's ground forces, as the second proposal would have us do, then targets might as well be attacked by cheap, short-range missiles controlled by cheap, short-range target acquisition instead of being attacked by much more costly systems from longer ranges.

It is not necessary to adopt the proposal in exclusive form. Instead of scrapping our present array of army divisions and air squadrons, as the Los Alamos group would have us do, we could simply add a layer of battlefield nuclear attrition capabilities to our present nonnuclear defenses, with the nuclear threshold remaining as high as the nonnuclear defense strength would allow.

In the past, in a climate in which it seemed feasible to deploy truly strong nonnuclear forces, the deliberate lowering of the nuclear threshold entailed by this proposal seemed unnecessary as well as dangerous. In the immediate present—that is, in the aftermath of the disheartening neutron bomb debacle—the inclination is to dismiss the proposal without giving it a hearing. But if the opinion is shared that our nonnuclear defenses are dangerously weak, the merits of the proposal must be examined in that respect.

Clearly its exclusive focus upon the immediate battlefield means that the proposal would have no decoupling effect if adopted. By failing to offer any protection whatever to anything behind the battlefield vulnerable to threat of long-range nuclear attack, it leaves open the entire question of how such protection is to be achieved. The proposed short-range missile force neither couples nor decouples vis-à-vis the Soviet Union, with its present array of forces and with the operational doctrine for the use of nuclear weapons prescribed by the manner and composition of that array of forces.

Nuclear Interdiction

Very actively discussed in U.S. professional circles, even though largely unpublished as yet, the nuclear interdiction proposal reflects a particular conception of the Soviet method of warfare as well as the implicit assessment that conventional air power is now of little operational worth for deep interdiction. This proposal calls for the deployment of large numbers of missiles with a range of up to 600 kilometers or so. With relatively low-yield warheads and highly accurate guidance (the exact form of guidance varies), these missiles would be targeted against Soviet second-echelon ground forces and their transport infra-structure. The operational premise, which is sound, is that the whole Soviet scheme of warfare critically depends on the maintenance of momentum in order to outpace defensive redeployments, which is essential to achieve the deep penetrations that lead to disruptive encirclements.

This essential momentum, it is argued, is generated in the Soviet method by the successive advent of fresh fighting echelons, which "shoot through" breakthrough points. This premise suffices to estab-lish the fact that if the additional fighting echelons are destroyed or even merely delayed, the entire Soviet scheme breaks down. This defines the primary target.

But a further premise is highly questionable: it assumes that the Soviet army's operational method is extremely primitive as far as the organization of a battle is concerned. The presumption is that the Soviet army would use a rigid and entirely preplanned steamroller method, in which the long columns of the advancing forces would be positioned on preselected axes, each axis being a battering ram with the head in the inter-German border and the other end deep in eastern Poland. On the basis of this presumption, it is argued that even if the difficulties of long-range target acquisition would reduce the capability of the proposed missile force to destroy Soviet maneuvering elements, an equivalent effect would be achieved anyway since the Soviet forces would have to disperse and would thereby be prevented from advanc-ing swiftly in good order. This further presumption is of course necessary because the acquisition of mobile targets so far behind enemy lines would indeed be very difficult to achieve quickly.

It is certainly true that in World War II, at least before 1944, the Soviet army was limited by its rigidity and its commitment to offensive thrusts on preselected axes, i.e., the steamroller method. But the Soviet army has changed greatly since then, and its battle organization to sustain momentum is now likely to be quite different. Instead of a linear battering ram (any swift advance of which would indeed be precluded by dispersal), NATO would probably be faced with a multi-plicity of initial thrusts, with each successful thrust receiving massive reinforcement. Instead of being rigidly aligned on preselected axes, the reinforcing divisions and regiments would be distributed fairly evenly over the territory, and each would be ready to move into some axis of advance opportunistically chosen by the higher headquarters. In this way, those initial thrusts that are successfully stopped become decep-tive, and those that happen to break through automatically become the main axes.

That method, which embodies the very essence of operational deception and thus makes nonsense of the frontal attrition tactics currently favored by the U.S. Army, undermines the military logic of the nuclear interdiction proposal. Far from being disrupted by an imposed dispersal, the method itself calls for a wide distribution of the regiments, each belonging to some theoretical reinforcement pool from which the chosen axes of advance could be sustained. Obviously, Soviet forces that were successfully attacked would be destroyed no matter what their intended method of deployment might be. But the imposed dispersal in itself would no longer yield much benefit, and without that element, the proposal loses much of its attraction. If only destruction can be achieved, why not exact such nuclear attrition at short range just behind the immediate battlefield, thus saving greatly both in the target acquisition apparatus and in the means of delivery?

On the other hand, it is clear that the proposed nuclear force would inherently offer a much broader deterrent protection than that envisaged in the first proposal. By having the capability to attack bases, airfields, depots, and so on as far away as 600 kilometers or so, the proposed missile force could persuasively threaten direct retaliation for any Soviet attacks upon similar NATO targets. Needless to say, this deterrent would still not be broad enough to remedy the shrinking of extended deterrence.

For the same reason, the proposed force of 600-kilometer missiles would also have no direct decoupling effect. In military terms, the main consequence of the proposal would be to replace the current, very costly, manned air interdiction capabilities with a missile force that was much cheaper and also much more reliable. Descending missile warheads are undistracted by air defenses and operate in all weather. This superior efficiency is attractive and so is the fact that the 600-kilometer missile force could be made less vulnerable to preemptive attack than the fixed-wing interdiction aircraft now in service. Even the high costs of a near real-time target acquisition apparatus could in fact be greatly moderated if platforms and sensors deployed for intelligence warning purposes could be exploited for this missile interdiction role also.

Unfortunately, replacing the current deep-interdiction aircraft with 600-kilometer missiles—the very feature that makes the proposal attractive in economic terms—would have a strong if indirect decoupling effect; 600-kilometer missiles would be replacing manned aircraft with far greater ranges, if only in one-way missions. This factor, in addition to the unreliability of the dispersal effect, is a powerful argument against the adoption of this proposal as now defined. Of course, an add-on program along similar lines, but drastically smaller, would be quite another matter.

European-Based Counterforce

Although the first two proposals are exquisitely American in character, notably in their emphasis on fighting rather than on deterrence, the third option of a European-based counterforce is somewhat more responsive to the European conception of the situation. Such a force would employ missiles with ranges greatly in excess of 600 kilometers, that is, weapons explicitly meant to reach targets deep within the boundaries of the USSR. They would not, however, be city targets; indeed, this option would offer no direct deterrent protection for European cities. Instead the goal would be to rehabilitate the deterrent protection of NATO military targets by threatening those of the Soviet Union.

NATO reinforcement access points, airbases, nuclear storage sites, and other high value installations are now vulnerable to attack by the new, accurate Soviet ballistic warheads and also by strike aircraft such as the Su-19. Given their own vulnerability, NATO strike aircraft do not themselves provide sufficient deterrence for this range, and the Pershing force is too limited in range. The proposed force, of accurate ballistic or, more likely, cruise missiles, would seek to deter this class of Soviet attack by threatening a directly equivalent retaliation. It is clear that this option would occupy a middle position in the spectrum of both deterrence and defense. It would not begin to substitute for the loss of extended deterrence, nor could it decisively remedy the inadequacies of the nonnuclear forces.

Given the basic geographic asymmetry between NATO and the Warsaw Pact countries and also the fundamental imbalance in nonnuclear capabilities, the proposed force could do little to improve the relative strength of NATO's defense, assuming a reciprocal exchange. NATO absolutely requires a positive contribution from air power, but the Soviet army could execute its own offensive operations with no air support at all; air-reinforcement entry points are obviously critical for NATO, but the Soviet Union can carry out reinforcement much more resiliently overland. On the other hand, by explicitly including targets within the Soviet Union, the European-based counterforce option begins to meet European political desiderata.

European-Based Countercity

Soviet ballistic missiles, in categories excluded from the peculiar definition of "strategic" affirmed in SALT, have long since made hostages of the cities of NATO Europe. That fact is the basis for the most dramatic, if perhaps the least significant, of the demands made upon extended deterrence. Now that the material basis of the latter is being eroded, it is being suggested by some that the United States should deploy MRBMs in Europe, to defend Europe in situ. It is interesting to note that the originators of the proposal were much more urgently motivated by a desire to protect SALT than by any strategic assessment of the European predicament. Realizing that only a perfect symmetry of obligations could make a mere parity of allowed capabilities acceptable over the long run, the inventors of the scheme thought that Europe-based MRBMs would dispose of the asymmetry created by the fact that the United States must protect allies in Europe exposed to a proximate Soviet threat.

If there were a choice between a highly reliable extended deterrence and the MRBM proposal, the former would of course be the much preferred alternative. Deterrence obtained by central systems from afar does not cost incremental money; does not attract counterforce attention to already weaker allies; and—above all—does not require political leadership, courage, and coherence when the deployment decisions must be made. But, in truth, there is no choice. Extended deterrence is now in an average condition: its scope is narrower than it was, but broader than it will be; its reliability is less than it was, but greater than it will be. In these circumstances, the MRBM option is not to be dismissed in spite of the motives of its originators, and despite its apparent decoupling effects. On the latter point, one may rest assured that with American-designed and American-controlled weapons aimed at Soviet cities, no decoupling need realistically be feared.

4

Detente Reconsidered (1976)

Viewed at close quarters, in the daily actualities of the unfolding process, the variegated phenomena of detente dissolve into three principal elements, two overt and much-publicized and one rather opaque. First, we have the public diplomatic events, from the elaborate ceremonies of reciprocal state visits to the coy opening formalities of negotiation where, until the doors are closed, images of guarded amity are carefully displayed before the photographers. Next we have the much less stately spectacle of the internal American debate which has attended the evolution of Dr. Kissinger's detente from the beginning. Always active, the debate undergoes periodic intensifications in a predictable rhythm, the two-four beat of congressional and presidential election years. Some might add that the debates do not merely become more intense but also undergo perceptible vulgarization in the proximity of these elections. Finally, there is the far more obscure process of Washington's bureaucratic politics, in which contending departments of government often seek to define the policies of detente by enlisting the support of unofficial pressure groups. Two of the three processes are American and only one is conjointly Soviet. There must of course be some counterpart in Moscow to all that goes on in America, but we are denied the privilege of knowing the hows and whys of Soviet decisions. To be sure, we have the brave, or at least relentless, Sovietologues among us who seek to divine who is for and who is against detente, or even particular aspects of the same. Those who struggle against the pervasive secrecy of Soviet political life do not deserve our scorn. But it must be admitted that their efforts have yielded meagre fruits, perhaps identifying a Shelest here or a Shelepin there as enemies of accommodation but with no ordered picture of the whole. Those who know how devious the forms of interoffice politics really are even in Washington must hesitate before presuming to understand those same politics as they are played in Moscow. Of Washington, it is by now well known that the principal opponents of

61

former American concessions in the Strategic Arms Limitation Talks are not in the Pentagon, and least of all in the Joint Chiefs of Staff, but rather in the offices of the Arms Control Agency. Who knows what permutations are wrought in Moscow? For all we know it may be the marshals of the Soviet army who support SALT and the foreign-policy specialists of the U.S. institute who oppose it. In short, we know nothing.

If therefore we discuss detente in this detailed perspective, we may speak only of the benefits and costs of its particular manifestations, from the SALT I agreement of 1972, to the Helsinki jamboree. Here each man may advance his own view, making his particular qualifications and stating his reservations in arriving at a particular verdict upon each agreement, or disagreement. We do know that the majority of American electors hold a generally negative view of all that has happened since 1972; we have this on the very best of authorities, the opinion samplers who are now serving the two candidates to the presidential office. Carter has discovered where the crowd is going and means to lead it by being suitably critical of Kissinger's handling of detente. Ford has seemingly discovered the same thing and he has carefully avoided making excessive reference to the late phases of detente over which he has at least nominally presided. Instead, Ford clearly prefers to stress his long-standing support for "national defense," and has attempted to depict Carter as a disarmer—with scant success so far.

But then if we lift our microscope from the daily course of events and view the processes of detente in a wider frame which spans the appointed lifetimes of American administrations, we might tend to discount the declamations of the current elections. Instead we would conclude that some sort of Soviet policy not easily distinguishable from Kissinger's detente has been pursued by each American administration since that of President Truman. And this would lead us to expect that postelection policies will be not too distant in substance from the preelection policies. However, there are some newly rediscovered ingredients in the situation which are of considerable importance. It is now evident that the Soviet Union's domestic and international conduct is still viewed by the dominant opinion in the United States as fundamentally immoral. And it has now been rediscovered that the legitimacy of the Soviet Union as now constituted is still disputed by the dominant opinion in the United States. Finally, it has now been rediscovered in Washington that the American body politic will not allow the president and his secretary of state to ignore its moral values, or for that matter its preferred mores as projected upon the international scene. I say "rediscovered" on all three counts because only four years ago an American president could calculate electoral advantage in embracing the leaders of the Soviet Union before the television cameras. It was only some time after the displays of affection, and after the signature of the first SALT accords that it became apparent that these acts were only viewed with favor in the very particular climate of a very particular time. Nowadays any advisor to the White House who might suggest a preelection Soviet summit spectacular would be recognized as incompetent, even in Ford's circle of advisors, whose repute for political wisdom is not high.

In the leader of the Soviet Union American opinion recognized in 1972 a potential partner in international tranquillity—a commodity then much in demand in a country far from tranquil. By contrast, only a few then advanced the notion that the leader of the Soviet Union was first of all its chief policeman, both at home and as far west as Prague. But as soon as the moment of American confusion passed, the images of conviviality were transmuted from those of a gathering of peacemakers to those of a gathering of ruthless and amoral political bosses, one dictatorial and one *aspirant*. As the Vietnam War descended below American horizons while Watergate loomed ever larger upon them, there was a concurrent amplification of the other view of the Soviet Union and its leaders. The intellectual dissidents, the Jews, and the doings of the KGB occupied the front pages day after day and week after week. Side by side with the latest scenes of the Watergate melodrama we had almost daily accounts of the struggles of unfreedom in Moscow. By 1974 it was all over, for the new-style detente barred the self-deception. Admittedly the self-deception was very pronounced: upon the advice of Kissinger, Ford gave his time to a soccer player from Brazil in preference to a writer from Russia; upon his advice also he went to Helsinki and Vladivostock. He is still paying for the first two; and he is only spared an electoral bill for the third because he was saved *in extremis* from finalizing a SAL accord. In 1976 we have thus returned to the starting point: only the less compromising forms of detente à la Eisenhower, Kennedy, and Johnson may be followed, because the American public will not allow its leaders to engage in close cooperation and intimate discourse with the leaders of the Soviet Union. The problem is hardly new: before 1917 also, American opinion would not allow the Czar's police methods and the Czar's pogroms to be left out of the diplomatic ledger. It may be lectured ad nauseam on the doctrines of internal sovereignty and on the realities of external power, but the American public will just not listen.

What sort of detente then is possible? First, as under Eisenhower, Kennedy, and Johnson there must be the generic self-restraint both in obvious and subtle forms which derives from that most obvious of all Russo-American imperatives, the avoidance of war. (It was at a particularly low point in the fortunes of detente and his own that Kissinger redefined his brand of detente as the avoidance of war *tout court,* as if there was anything new in that.) Next, there must be the earnest pursuit of agreement on the limitation of those armaments and acts which are particularly unpleasant, while being also of marginal military utility; the classic case being of course the Outer Space Treaty, which prohibits nuclear detonations outside the atmosphere. Third, there must be a steady readiness to engage in a substantive diplomatic dialogue with the Soviet Union, so long as this in no manner entails the legitimization of phenomena to which the American public will not in fact accord legitimacy; here the classic case is the Soviet Union's domination of Eastern Europe. What then does this old-new detente exclude? Principally those American policies which strengthen the Soviet Union and its regime economically and therefore politically from the extension of government credits at concessionary rates to the transfer of technology. And then it is also demanded that American leaders avoid manifestations of excessive conviviality with the leaders of the Soviet Union, for good and moral men should not consort with the ultimate heads of the KGB. Here the image to be avoided is that of Nixon winning and dining in the Kremlin while Jews were being arrested just outside its walls.

Albeit, at arms' length, much can therefore continue as before. Except of course for the economic and technological meta-processes which loomed so large in the visions of detente à la Russe. American farmers will remain free to sell their wheat to all buyers but IBM will have to be more circumspect in its marketing. And of course the multibillion credits that were to finance Soviet economic growth are simply not on.

It is important to recognize that the Soviet Union has no effective sanctions to apply, unless extreme. In particular, it may not discourage this reversion to detente in the old mode by withholding benefits arising from the post-1972 version of the same. Having notably failed to limit the pace of its armament build-up, having conspicuously failed to show restraint in the Middle East (in 1973) and in Angola most recently, and many a point in between, it may not now threaten to increase its armaments, or its activism. For both have been maximal in any case. Least effective would be to threaten a further acceleration of the strategic competition: Soviet missile factories are already working at full capacity—and all for nought, if the goal is a meaningful supremacy. For a fraction of the cost to the Soviet Union, the United States can match every increment in Soviet strategic power with one of its own. Arms limitation agreements which control specified arms, and established superiorities can matter little in any case: the dynamics of the competition are *primarily* manifest in weapon innovation, which brings forth new capabilities that quickly offset the old.

Do we then leave it at that, with this modest vision of an unexciting but not unproductive detente? Not quite. Not, that is, if we take a further step back from the present and widen our perspective from months to years to decades and centuries. In a historical view the processes of detente post-1953, as the processes of detente à la Kissinger, appear as two more episodes in the long struggle of Europe for the inner core of Russian public life. They are two more episodes in the unfinished saga of the Europeanization of the Russian people. A process seemingly firm in its direction until 1917, the Europeanization of Russia was thereafter interrupted by revolution, civil war, oppression, more war, and a final reversion to oriental despotism of the last years of Stalin. Even now, it is worth restating that in 1914 the Russians were industrializing quite rapidly and their political life was being modified by European humanism, legality, and the pluralism that both stimulate, and which stimulates both. If not for these two things, there would have been no industrial workers in Petrograd to establish the prerequisites of revolution and there would have been no revolutionaries to lead them, men released from an easy detention by the incipient legal restraints which already moderated the scope of the Czar's police oppression. The revolution of Kerensky was universally seen as a giant step in the Europeanization of Russia, and indeed it was. There was every reason to believe that pluralist humanism, the rule of law and the ordered decentralization of power would finally settle in the lands of the Russians. Even the *coup d'état* of Lenin was not seen as a retrograde step; after all it was Lenin himself who accorded in his writings the right of self-determination to the peoples dominated by the Russians, and it was Lenin who acquiesced in the independence of the Finns and the Baltic peoples. Self-determination was not then universally an essential part of the European ideal, except for Europe itself (the English wept for Italian patriots, but could imprison those of India). But it was recognized that in the case of the Russians there was an organic nexus between domestic despotism and Great Russian colonialism. So long as the Russian empire remained the prison house of peoples, the Russians themselves would not merely be the prison-guards but also prisoners themselves. Then came civil war and Lenin's Swiss-bred Europeanism was washed away to reveal the Asiatic despot who rebuilt the Czarist police-state in a far bloodier form. Thereafter the relapse proceeded.

Since 1953, there has been a major reversal, and we have seen great changes in the Soviet Union. These changes are perhaps best exemplified by the phenomenon of dissidence itself, which already reveals a relative freedom and a relative legality in the contemporary Russian empire. Dissidence is of course manifest in its complaints against oppression and illegality, but it is perfectly clear that it could not exist but for a modicum of their opposites. We must do everything we can to encourage the further Europeanization of the Russians. To the extent that they become more European, it is more likely that the Russians will once more contribute to the culture, as they once did before the tragic silence which supervened under Stalin forty years ago. To the extent that the Russians become more European, it is more likely that they will contribute to science as they once did before the sources of creativity were strangled by the dictator in the particularly ill-fitting robes of the Scientist-King. To the extent that the Russians become more European, it is more likely that they will begin to contribute in the economic sphere as well, becoming industrial innovators rather than mere suppliers of raw materials.

Only in the perspective of this secular and uncertain process does the often unreciprocated concessionary policy of Kissinger and his colleagues acquire meaning, and substantial justification. Concluding in *media res* one more thing must nevertheless be said, in case it is not evident already: in encouraging the Europeanization of Russia we are not imposing our own ideals on the Russian people. We are merely allying ourselves with a particular segment of this great and tragic nation against another. In Russia there have always been those who believed in humanism and legality alongside the more powerful body of the supporters of despotism and obscurantism. The two strains were in evidence in the life of the Russian Church from the conversion until the end; the two strains were in evidence under the Czars and among Lenin's associates, and even within Lenin himself. The two strains are in evidence now, and not all believers in humanism and legality are to be found among the dissidents, nor are all believers in atavistic reaction, despotism and oppression to be found in the ranks of the Soviet government.

5

Military Balance
and Deus Ex Missiles (1983)

The Lebanon War

As the fog of war dissipates, it is becoming more and more obvious that Israel's victory over the leading Soviet allies in the Middle East, with the subsequent demonstration of the impotence of the Arab "radicals," and—above all—of the Soviet Union's inability to protect its clients, has done a great service for the United States and Western interests as a whole. (Whether Israel was wise in paying the price in blood, treasure and international repute for the benefit of powers determined to deny any gratitude whatsoever, is another matter.) But the Lebanon war could eventually turn out to be something of a disaster if the results of the fighting are interpreted to mean that Soviet weapons are grossly inferior to American arms, thus giving rise to dangerous delusions about the military balance in Europe and worldwide.

The Israeli-Syrian fighting in Lebanon during June unfolded in three distinct operational settings: the air-to-air war; the Israeli attack upon Syrian air defenses; and the war on the ground, primarily fought between armored forces. Each was largely self-contained. As it happens, each of the three has a very different meaning for the military balance in Europe and beyond, and for our own defense policy.

The Air War

The results of the air war were certainly unequivocal: contrary to some initial reports, the Israelis lost *no* aircraft in the air battles of June 7-10, in which American-built F-16s and F-15s as well as Israeli Kfirs shot down at least 85 (and very likely more) Syrian fighters of the MiG-21 and Mig-23 type (each has several variants). There was of course nothing new in the fact that the Israelis prevailed in air combat, but in the past the ratios were roughly 20-30 to one, not 85-95 to zero as in the June battles. One new factor responsible for eliminating the marginal Israeli losses of the past was the use of the E-2C airborne radar aircraft. In previous air battles, Israeli fighters absorbed in dog-fights were sometimes caught by surprise and "bounced" by additional Syrian fighters which arrived on the scene unobserved. With the E-2C, however, the Israelis had due warning, and could send into action uncommitted fighters, held ready at altitude to engage the newcomers. Only the United States Navy (and Japan) have E-2Cs, but the American air force and the North Atlantic Treaty Organization (NATO) as a whole have the much larger AWACS aircraft which should perform a similar "battle management" role. Great claims have been made for the airborne-radar aircraft but their importance should not be over-estimated, if only because "management" is of small avail if there are not enough ready fighters to exploit the information.

The larger and actually more decisive question is of course the quality of the fighter-aircraft themselves. For two decades after 1945, the United States fighters were so superior to their Soviet counterparts that any numerical comparisons had little meaning. Then, during the 1960s, the quality gap diminished quickly, with powerful implications across the board—since within the overall American strategy the weakness of ground forces is supposed to be offset by large scale air support, which in turn requires the prior achievement of air superiority, mainly by fighter combat. Now once again, it is clear that the latest, first-line American fighters are greatly superior to their Soviet counterparts. But since these fighters would be greatly outnumbered in a large European war, it is important to estimate the magnitude of their superiority. In other words, we must subtract the pilot factor from the 85-0 equation to arrive at a valid estimate of the aircraft themselves. Israeli pilots are the products of a very unusual system (we educate future generals, who can also fight in their younger years; the Israelis train fighter pilots, a few of whom are then retrained for subsequent command in an air force with hundreds of aircraft and a mere handful of generals), and it is no secret that they are very good, not merely superior to Syrian pilots but also to those of NATO.

But even if we make a large allowance for pilot quality, it is clear that our new fighters are superior to their current Soviet counterparts. This, however, means much less than we would like for the European air balance, since we cannot properly compare Soviet fighters in mass deployment with our latest and best, but must instead compare them with their true equivalent, that is, NATO fighters also in mass deployment. And when we do that, the comparison is by no means favorable: NATO air forces are still largely equipped with 1960s aircraft (F-4 Phantoms, older F-104s, slow Jaguars, light Mirages and so on), the products of the 1950s development lag brought about by the emphasis of those years on "massive retaliation" and all-nuclear war. Taking the pilot factor once more into account, NATO should still prevail in unit quality, but this is in turn offset by the numerical advantage of the Soviet air force. This, by the way, justifies the claim of the "military reformers" that we should add numbers to quality, by also producing light and simple MIG-21 type fighters alongisde the F-15s and F-16s, which will always be in short supply.

The Israeli Attack on Syrian Air Defenses

When we examine the second war that was fought by the Israelis against the Syrian air defenses in the Beka'a, we find much less direct relevance for the European balance and a far stronger lesson for our defense policy. The Syrian deployment was huge, including 19 missile batteries and many guns, and it was defeated by what was in effect a large-scale electronic commando operation based on refined intelligence, split-second timing and the extensive use of deception. The standard tactic—which no longer works—relies on air-launched missiles that come in on the beam of the enemy radars; the Israelis, it seems, turned that tactic upside down, by using mere decoys in that role, while attacking for real with other weapons that were coming in too low for radar detection. And this was merely one deception among many, in a very elaborate scheme which out-maneuvered, rather than out-gunned the Syrian system. Again, the Israelis lost no aircraft in the process, a result all the more remarkable for the fact that they had lost the advantage of surprise since the strike came on the third day of war. NATO could not hope to reproduce this result. In the first place, the Syrian air defenses were grouped within easy reach of Israel, whereas in Europe the overlapping Soviet missile belts reach across the full width of East Germany, Poland and the Soviet Union itself. This means that ground rockets and other shorter-range weapons could not be employed as they were in the Beka'a. Second, the Israelis used a great deal of equipment that NATO lacks, and which is in short supply even in our own forces.

For example, the Israelis relied heavily on remotely-piloted vehicles (RPVs), basically large model aircraft, to collect intelligence and to confuse the air defenses. At present, there are *no* operational RPVs in our forces, largely because pilot-dominated military bureaucracies have been most ingenious in finding reasons to do their job with manned aircraft. Generally, one lesson of the episode is that we too should overcome the bias against electronic warfare that favors shiny new aircraft over ugly little black boxes. Once the worst offender, the Israeli air force learned that lesson in 1973 at great cost; we now have a chance to do so for free.

It would be a tragic misunderstanding of what happened in the Beka'a to think that the panoply of Soviet air defenses is ineffectual. In a straight "force-on force" attack of the sort that NATO would have to mount in the urgency of war (with outnumbered ground forces pleading for immediate air support), the high-altitude SAM-2s and SAM-5s, and the medium-altidue SAM-3s and SAM-4s would achieve few kills but force NATO aircraft to fly low, and then the SAM-6s and shorter-ranged SAM-8s would achieve many kills, with still more going to the many guns and smaller SAM-7 and SAM-9 missiles.

The Ground War

The ground war was a simpler affair: fighting for the first time under the command of an army corps, five powerful Israeli armored divisions defeated the somewhat smaller and clearly outgunned Syrian forces. The reason why this too was a remarkable achievement was because of the terrain: the narrow single-lane roads, often twisting on the slopes of high mountains, should have meant a very slow operation, whereas the destruction of the Syrian front was achieved in a day, a night and a morning (June 9-10). Towards the end of the fighting, the Syrians sent in their 3rd Tank Division, equipped with T-72s, thereby achieving the destruction of those tanks, and the creation of a myth. This holds that the T-72 is a "piece of junk" which means (1) that the tank-centered Soviet army is weak, and (2) that we ourselves do not need the new and costly M-1 tank.

With thousands already deployed in the Soviet army, the T-72 is in fact the generational equivalent of our late-model M-60 Pattons, and not the M-1 which has just entered production and which will not be in mass deployment for several years. In due course, when we know more about it, we should compare the new Soviet T-80 tank with the M-1. In the meantime, the only valid comparison is between the *deployed* tanks, our improved M-60s versus the T-72s. Now it is known that the Israelis have many Pattons and it is also known that they destroyed a number of T-72s; in Europe, too, M-60s would be fighting T-72s, but the syllogism does not hold: the Israelis certainly did not rely on M-60s to defeat the T-72s.

Our tank modernization efforts have been so slow that the M-60 has the distinction of being *the* standard American weapon in its category while also being definitely obsolete, except in regard to its very modern electronic fire-control device. The Israelis are glad to have M-60s to make up numbers, but consider it the least capable of all their tanks, in third place after the Merkavas and the British-made Centurions; over the years, they have added all sorts of improvements but the basic design of the M-60 has irremediable defects of design that make it inferior to the Centurions, which have also been modernized. All these tanks, like our own, have 105mm guns, as compared to the 125mm guns of the T-72, and the latter in addition has a much stronger armor envelope than the M-60 or the Centurion (but is less protected than the Merkava). The Israelis are confident of being able to defeat T-72s, even if outnumbered unlike last June, but would dread the prospect of having to fight greatly superior numbers of T-72s with M-60s alone, as the United States forces in Europe would now have to do. Moreover, it turns out that the Israelis defeated the very strong armor of the T-72s with a new type of ammunition (M-III) which they produce themselves. Our current 105mm tank ammunition could not penetrate the T-72's frontal armor at all, except at dangerously short ranges in which the sheer size superiority of the Soviet 125mm gun would be devastating. (As usual, we lack an improved ammunition of the Israeli kind because the army is trying to produce a yet more advanced type, leaving its forces with a grossly inferior 105mm round in the meanwhile.)

The war in Lebanon is replete with lessons, though mainly about leadership, tactics, and strategy rather than weapons technology. Serious analysis could yield great benefits for our defense policy, but only if we avoid crude misinterpretations of the evidence based on false comparisons between the best of Western equipment in the best of hands, and Soviet equipment sometimes less advanced but always present in far larger numbers.

Deus Ex Missiles

Antiaircraft missiles, antiship missiles, antitank missiles, antiradar missiles, and even antimissile missiles have been around for years now, but it is only quite recently that we have seen the emergence of the antidefense missile, aimed primarily at the defense budget. We are hearing a great deal about the virtues of antitank missiles: they cost only a few thousand dollars apiece and have what the trade calls high-kill probability. This being so, why should we spend nearly $3 million apiece for the M-1 tanks that the army wants so badly, particularly as our opponents in combat have their own antitank missiles? The sinking of the HMS Sheffield in the Falklands war revived the naval version of this argument. Why build such expensive ships, costing hundreds of millions and even billions of dollars, when they can be sunk by "cheap" missiles costing only hundreds of thousands of dollars? In both versions the services are pictured as trying to foist upon us some sort of antiquated armored or maritime cavalry that will not only fail us in our hour of need, but will squander our treasure as well.

It goes without saying that all institutions are conservative, that they will strive to perpetuate existing forms. And military institutions, with their rigid hierarchic structure, are even more conservative. Their tendency to reject the new should be taken for granted, especially when the old embodies the ethos of the institution, and the new makes its appearance in the guise of mere functionalism. The army defended its beautiful horses against the ugly and soulless tank—until the tank too developed its own ethos, which now in turn evokes powerful loyalties against the sterile functionalism of the missile.

Since military conservatism is an inherent tendency, since it can be costly and may also place us in great danger, we should indeed be ready to intervene and impose change when it is needed. But perhaps nowadays the greater danger is the opposite: the tendency, at least in our society, to impose the "modern" before it can really do the job, and sometimes when it will not work at all. In our culture especially the "tech-fix" illusion is deep and wide, and it affects our armed forces as much as any other American institution. In Vietnam, "tech-fix" delusions were ruthlessly imposed on a mediocre military leadership by Robert McNamara; systems analysis and statistical body-counts displaced strategy, "manpower management" displaced leadership, and the serving of hi-tech equipment (from multisensor barriers to "people sniffers") displaced tactics and the operational art of war. By such devices we compounded our defeat by costly and demoralizing absurdities.

A proper suspicion of military conservatism must therefore be balanced by a prudent attitude toward technological wizardry. In considering the true military worth of the antiship missile, for example, the first thing to note is that this is the third apparition of that weapon. In September 1943, the German *Luftwaffe* sunk the Italian battleship *Roma* and seriously damaged the *Italia* with the Fritz-X, the first effective air-launched antiship missile; since the Germans were about to deploy the even more effective HS-293 missile, the landings at Salerno, Anzio, and in Normandy ought logically to have resulted in huge losses of Allied shipping to those precursors of the Exocet. But in fact the Fritz-X was out of business by the end of 1943 and the HS-293 was withdrawn a few months later. Powerful indeed against warships that were unprepared for them, both missiles were easily neutralized by simple countermeasures—devices as "cheap" as the missiles themselves. (The Fritz-X and the HS-293 were radio-guided, gliding missiles, and these two qualities proved their undoing. The gliding required that the missile be launched from a particular distance from the target; once that envelope had been identified, Allied fighters turned it into a death trap for German planes. The radio-guidance provided the opportunity of jamming; once the frequencies were located and devices installed, the missiles quietly glided down into the sea.) The second apparition was all the more dramatic since the background was uncrowded by the events of a large war: on October 21, 1967, at a time when a cease-fire was in effect, the Israeli destroyer *Elat* was sunk by Styx missiles launched by Egyptian missile boats. Articles and editorials very much like those being printed nowadays proclaimed the virtues of Styx missiles and the hopeless vulnerability of all warships. But in fact the Styx was soon thereafter neutralized by relatively simple countermeasures. Six years after the sinking of the *Elat*, more than fifty Styx missiles were launched at Israeli ships during the October War: only one suffered very slight damage from a near-miss, and none at all were sunk.

Now that the Exocet has had its turn, we can anticipate two parallel developments: ample sales of that missile, and concurrently, the swift development of countermeasures that will do the Exocet what was done to the Fritz-X and the Styx. For there is one rule in strategy that is truly universal: nothing worthwhile comes cheaply. There are *no* cheap weapons that cannot be defeated by equally cheap countermeasures. The antiship missile is cheap because it is a single-purpose weapon; the warship is expensive because it had broad capabilities— and one way of using the space and weight, the sensors, power sources, and crew that a warship has to offer is to equip it with countermeasures against the antiship missile.

Naturally, the defender does not get off cheaply either: besides the direct cost of jamming devices, antimissile weapons, and so on, there is a definite loss of offensive capability for every addition to a ship's defenses. Eventually the point will be reached when the procedure will no longer be worthwhile, when defenses against missiles and other weapons will take up so much of the warship's capacity that it will become more trouble than it is worth. But it will not be a "cheap" missile such as the Exocet that will put the surface warship out of business but some more versatile and far more costly weapon—still, for now, a distant prospect. To be sure, American carrier task forces must already contend with a real Soviet missile threat. But that threat comes not from a few cheap missiles but from hundreds of rather costly weapons, launched from dozens of still more costly nuclear-powered submarines and naval bombers—and so far the balance is still favorable to the carrier forces, if properly equipped and trained (unless nuclear weapons are used against them—but then, of course, land bases would be even more vulnerable). The point is that at this interim stage in the evolution of naval weapons, it would be as foolish to abandon the large warship as to fail to provide it with proper defenses against antiship weapons, old and new.

The antitank argument is an old story, but it has recently acquired an entirely new and larger dimension because it has been invoked by the proponents of a "no-first-use" nuclear policy for NATO. The Soviet Army consists largely of armored forces, organized, equipped, and trained to carry out deep penetrations at high speed, in a modern and more powerful version of the blitzkrieg. NATO's declared readiness to use tactical nuclear weapons is meant to deter a Soviet armored offensive which could not reliably be defeated by the alliance's nonnuclear forces, if only because the latter are widely distributed while the Soviet Army could concentrate its divisions in a few powerful mailed fists to mount a surprise offensive. (And if a gradually escalating crisis provided enough warning for NATO to reinforce and deploy properly, this would be offset by the fact that the Soviet Union could mobilize many more divisions.)

That is the setting in which the humble antitank missile now enters, stage left, as a strategic solution to NATO's problem: it is no longer necessary to deter a (nonnuclear) Soviet offensive by tactical nuclear means, because with enough antitank missiles NATO forces could defeat a Soviet blitzkrieg. There is no doubt whatever that today's antitank missiles can indeed destroy tanks. It is also true that the cost ratio between a modern tank and the best antitank missile system, the TOW, is roughly one hundred to one, and even the cheapest armored vehicle costs several times as much as the TOW. Thus NATO could certainly deploy many antitank missiles for each Soviet tank, especially if it gave up its own costly tanks. With hard facts like these on their side, it is not surprising that so many arms-control advocates, Members of Congress, and journalists have concluded that the tank in particular and armored forces in general are today's cavalry, perpetuated only by bureaucratic sentimentalism.

But strategy is more complicated than bookkeeping, because it has multiple dimensions. In a technical context, when the antitank missile is tested on the proving ground against a target of armor plate, the missile wins; but in a tactical context, where the missile crew must operate under enermy fire, or are unable to exploit the weapon's long range because of intervening obstacles or just smoke, the high-kill probabilities theoretically achievable are drastically reduced. In an operational context, where the total forces of both sides enter the picture, the antitank missile is finally reduced from a panacea to the more modest role of an adjunct to older antitank measures, including mines, terrain obstacles, antitank guns, and above all other tanks—the last having the huge advantage of being able to move about under fire to do their job. But it is at the level of theater strategy that the delusion is finally exposed, for then we find that the cheap and abundant antitank missile will in fact be grossly outnumbered by the costly armored vehicle, for the latter can be concentrated in narrow-deep columns while the antitank missiles must be distributed all across the frontage established by the chosen NATO strategy, thus being immobile if dug in, or vulnerable to artillery if not. Of course, the antitank missile can also be made mobile, indeed, much more mobile than any armored vehicle, by the simple expedient of providing a helicopter to carry the missiles and its firing crew—but then, of course, the weapon will no longer be cheap, and will in fact cost many times as much as the average armored vehicle. If we are more modest, and merely seek to equal the mobility of armored forces in order to be able to match enemy concentrations with our own, we must provide a cross-country vehicle for our missile, and then armor to protect its crew. Next we would find that the missile-firing armored vehicle will usually be inferior to gun-armored vehicles: guns have a much higher rate of fire and, moreover, the new "Chobham" armor of the latest tanks is very effective in stopping the hollow-charge warheads of missiles, while tank guns rely primarily on high-velocity shots that destroy armor by plain kinetic energy—and the new armor is as vulnerable to them as the old. Having wrapped our antitank missile in armor and given it cross-country mobility, we would then find it advantageous to replace or supplement the missile launcher with a gun—and we would thus reinvent the tank.

Part II

The Uses of Naval Power

6

The Political Uses of Sea Power:
The Theory of Suasion

In having a peacetime political function in addition to their combat capabilities, naval forces are like all other forms of military power, only more so. The familiar attributes of an oceanic navy—inherent mobility, tactical flexibility, and a wide geographic reach—render it peculiarly useful as an instrument of policy even in the absence of hostilities. Land-based forces, whether ground or air, can also be deployed in a manner calculated to encourage friends and coerce enemies, but only within the narrow constraints of insertion feasibility, and with inherently greater risks, since the land nexus can convert any significant deployment into a *political* commitment, with all the rigidities that this implies.

In wartime, the political uses of sea power are naturally relegated to the background in the formulation of naval strategy, which concentrates on combat capabilities, i.e., "sea control" and "projection," to use the current jargon of the U.S. Navy.[1] In the absence of general hostilities, however, a reverse priority applies, and though the prolonged confrontation of the Cold War has retarded the process, the focus of Great Power naval strategy has been shifting to missions that are "political" in the sense that their workings rely on the reactions of others, and these are reactions that naval deployments may evoke, but cannot directly induce.

In order to evaluate the political impact of naval deployments and assess their political utility, the distinct modes in which any political effects are generated must be defined and classified, just as the combat capabilities of a fleet are assessed by computing the different tactical and strategic capabilities that are found within it. Just as such an assessment is based on a previous classification of the various functional missions (antisubmarine, antishipping, interdiction on land), a political evaluation requires its own system of analytical classifications, even if the precision and stability of any definition in this area must inevitably be limited. A set of definitions has accordingly been presented below to avoid the use of terms that have already acquired unwanted and possibly misleading connotations.[2] These definitions are based on the conveniently neutral term *suasion,* whose own meaning usefully suggests the indirectness of any political application of naval force.

Armed Suasion in General

If one tries to disentangle the role of the U.S. Sixth Fleet in a major international event such as the 1972 expulsion/withdrawal of Soviet forces from Egypt, one would have to begin by isolating the third-party and adversary reactions which the Sixth Fleet may reasonably be deemed to have evoked. Some are obvious. First, by virtue of its perceived capabilities, by its role as a major asset and symbol of the United States, and through the intermittent manifestation of political will in Washington with respect to its possible use, the Sixth Fleet foreclosed a number of military options otherwise open to the Soviet leadership. This represents *latent* suasion in a *deterrent* mode at various levels of intensity. These Soviet options obviously included a range of offensive moves with respect to Israel, and the Sixth Fleet thus precluded Soviet actions that could have enhanced the alliance-worth of the Soviet Union in Egyptian eyes. This in itself can be deemed to have been one of the basic causes of the event to be explained.

A second effect was also latent but supportive rather than deterrent: Turkey, a vulnerable ally in control of the Straits, was given tacit support by the presence of the Sixth Fleet. This tended to strengthen Turkish resolve to continue to enforce the Montreux Convention rules which affect Soviet use of the Straits; and of course these rules would prejudice the security of any Soviet naval, or sea-supported, forces on the far side of the Straits in the event of war.

A third effect, also latent and supportive, was felt by the Israelis. While the Sixth Fleet strengthened Israeli resolve to resist Soviet pressures and threats (by providing an "insurance" element in their calculations), its presence also militated against Israeli activism, since the net incentive of a preemptive attack on the Soviet forces in the area was much reduced by the ultimate security that the Sixth Fleet was thought to offer. As a result, the Fleet was instrumental both in encouraging Israeli resistance to the Soviets (which devalued the worth of the Soviet connection to Egypt), and in preventing a very dangerous Israeli move, a move that could not have been excluded from the realm of possibility given the open-ended and ominous nature of the Soviet build-up in Egypt in 1970.

A fourth, one-time effect, manifest at an earlier point in time, was an example of *active* suasion, both *coercive* and *supportive:* the declaratory intervention of the United States in the 1970 Syrian-Jordanian crisis. In this instance, there was both a negative (i.e., deterrent) mode of coercion, in that the United States sought to prevent a Soviet intervention, and a positive ("compellent") mode, in that the United States demanded that Syria withdraw its armor from Jordan, for which the Palestinian cover was too thin to be plausible. The supportive element of this instance of active suasion was of secondary importance, because reports indicate that the Jordanian leadership was already fully resolved to resist the Syrian attack; nevertheless, the insurance provided by the "projection" capabilities of the Sixth Fleet must have intruded on Jordanian calculations by reducing any incentive to seek a political settlement and a fresh compromise with the Palestinian military organizations. What makes this one-time application of sea power *active* is the nature of the operation: it involved specific, though not necessarily overt, American demands. The fact that the operation also involved redeployment and reinforcement of the Sixth Fleet is not, on the other hand, fundamental to our distinction between *latent* and *active* suasion.

Whether it is active or latent, coercive or supportive, deterrent or positive (i.e., "compellent"), *armed suasion* is manifest only in others' reactions; it is the general concept that embraces all such reactive efforts. Any instrument of military power that can be used to inflict damage upon an adversary, physically limit his freedom of action, or reveal his intentions[3] may also affect his conduct, and that of any interested third parties, even if force is never actually used. The necessary (but by no means sufficient) condition is that the parties concerned *perceive* (correctly or otherwise) the capabilities deployed, thus allowing these capabilities to intrude on their view of the policy environment and so affect their decisions. Armed suasion, therefore, applies to others' reactions, and not the actions, or the intent, of the deploying party. The latter may exercise suasion in order to evoke certain reactions, but cannot achieve them directly, as combat effects can be achieved by the application of force.

Because suasion can only operate through the filters of others' perceptions, the exercise of suasion is inherently unpredictable in its results. Routine fleet movements which were not intended to pose a threat may be seen by others as threatening (since the threat is *latent* in

TABLE 1

The Political Application of Naval Power: A Typology

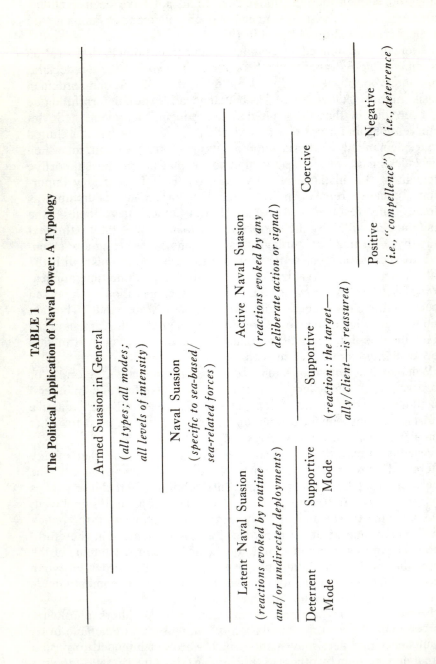

Armed Suasion in General

(all types; all modes; all levels of intensity)

Naval Suasion

(specific to sea-based/ sea-related forces)

Latent Naval Suasion

(reactions evoked by routine and/or undirected deployments)

Active Naval Suasion

(reactions evoked by any deliberate action or signal)

Supportive Mode

Deterrent Mode

Supportive

(reaction: the target— ally/client—is reassured)

Coercive

Positive

(i.e., "compellence")

Negative

(i.e., deterrence)

the forces themselves). On the other hand, a deliberate but tacit threat may be ignored, or worse, may evoke contrary reactions. In addition, in the decision-making arena of the target state, the threat (or supportive) perceptions evoked by the forces deployed have to compete with all other political pressures that have a bearing on the decision, and the final outcome of this complex interaction is impossible to predict.

The exercise of armed suasion in peacetime need not exclude the actual use of force, where the use is "symbolic." Because this is the era of undeclared conflicts brought about by the new and significant inhibitions to the overt initiation of war that have become manifest since 1945—the term "peacetime" now defines only the absence of *general* hostilities conducted at a *high* level of intensity. It follows that no firm dividing line can be established between the use of threats and the actual infliction of damage albeit in small doses. As long as the purpose and context of the use of force remains political, i.e., intended to evoke suasion effects rather than to destroy enemy forces or values, it cannot be arbitrarily excluded from the range of political instrumentalities provided by naval forces in "peacetime." But the political use of symbolic forces *does* require that the target state recognize its symbolic nature, i.e., that the damage inflicted has been *deliberately* minimized. This in turn requires the deploying state to discriminate successfully between what is and what is not symbolic in terms of others' perceptions, which may be quite different from its own. The same bombing raid by ten F-4s may be seen in Cairo as the beginning of a large-scale attack; in Washington and Hanoi, as purely symbolic; and in, say, Damascus as an all-out air offensive. In other words, the use of symbolic force as a way of augmenting coercive suasion entails an additional element of uncertainty.

Nor is the exercise of suasion predicated on the absence of hostilities, even general hostilities. During World War I, *tactical* suasion kept the High Seas Fleet well clear of Atlantic shipping lanes; it also kept units of the Grand Fleet from bombarding U-boat bases on the Flemish coast. The former case of tactical suasion was based on German *perceptions* of British naval power that may or may not have included a full appreciation of the vulnerability of the Grand Fleet, including its defective large-caliber shells and the weak horizontal armor of its battle-cruisers. Similarly, the deterrence evoked by Western anti-submarine warfare forces may continue to operate on the Soviet submarine force, even if the increasingly unlikely prospect of a Soviet offensive in Western Europe were to materialize. In the past, such deterrence may have tacitly discouraged the Soviets from exercising the option of a submarine campaign against Western shipping, and such deterrence would not cease merely because of the outbreak of hostilities. If peace is divisible, so is deterrence.

Because armed suasion operates on both the tactical and the political level, contradictions between the two may occur, thus presenting very serious decision-making problems. One may readily visualize a situation in which the tactical suasion of, say, greatly reinforced Soviet naval and naval-air forces based in Egypt and Syria could discourage the U.S. Navy from deploying its own forces in the eastern Mediterranean for reasons of elementary military prudence, while at the same time at the political level it may be thought that the worth of continued deployment was greater than ever. Political considerations would normally be of overriding importance in the decision-making arena, but only if the issue reaches the political level. If all concerned, on all sides, share the same perception of the balance of forces, the problem should not arise, since unviable forces would be discounted as either threats or supporting elements in the power equation (except in extreme cases where military forces are made hostages to commitments). More commonly, however, the same set of rival forces is evaluated quite differently in different quarters so that given our hypothetical Soviet build-up, for example, the Sixth Fleet could at the same time be very vulnerable and yet still seen as very useful in the political arena. Since men at the tactical and political levels have quite different responsibilities, contradictions between the two levels of suasion can be a source of acute internal controversy, just as the conflict between tactical and political priorities has been a chronic source of tension between soldiers and politicians in times of war.

To summarize, *armed suasion* defines all reactions, political or tactical, elicited by all parties—allies, adversaries, or neutrals—to the existence, display, manipulation, or symbolic use of any instrument of military power, whether or not such reactions reflect any deliberate intent of the deploying party. *Naval suasion* refers to effects evoked by sea-based or sea-related forces.

Latent Suasion

Effects evoked by the deliberate exercise of armed suasion where the intention is to elicit a given reaction from a specified party are defined as "active" in what follows. As against this, the undirected, and hence possibly unintended, reactions evoked by naval deployments maintained on a routine basis are defined as *latent*. Latent naval suasion continuously shapes the military dimension of the total environment which policymakers perceive and within which they operate.

By those who perceive them, the specific capabilities deployed are seen as potential threats or potential sources of support. As such, they influence the behavior of those who deem themselves to be within reach of the forces concerned.

In the deterrent mode, the range of capabilities perceived sets a series of tacit limits on the actions that may otherwise have been considered desirable or, at any rate, feasible. In this sense, one should not speak of a presence so much as of a shadow that impinges on the freedom of action of adversaries, because the capabilities perceived can be activated at any time, while the formulation of the intent to use them can be both silent and immediate. The ultimate readiness to resort to force is, of course, indispensable; without it there can be no armed suasion whether latent or of any other type.[4] It is therefore misleading to make any dichotomy between "peacetime presence" and "wartime" combat capabilities, since a "presence" can have no significant effect in the absence of any possibility that the transition to war will be made. Latent suasion is therefore the most general (in terms of intensity) and geographically the most widespread form of deterrence; as such, it is likely to be the most important class of benefits generated by sea power. More will be said about deterrence as such below. Here is it only necessary to note the inherently tacit nature of latent deterrence and its major implications: first, the wide scope of miscalculation that the tacit nature of this form of deterrence inevitably implies; second, the greater flexibility of this form of deterrence, in that in the event of a failure there is no compulsion to carry out an act of retaliation that was never threatened overtly; third, the corresponding weakening of deterrence where no rigid commitment to implement retaliation obtains. To speak softly while carrying a big stick may be less effective as a deterrent than to make a firm, overt commitment to use a rather smaller stick.

The second mode of latent suasion is supportive. The deployment of naval forces is a continuous reminder to allies and clients of the capabilities that can be brought to their aid. Moreover, with its ready intervention potential, a fleet can give a tangible content to any prior commitments that may have been made. Normally, the effects of this mode of suasion are seen as solely beneficial: allies are encouraged to adhere to alliance policies and dissuaded from conciliating adversaries at the expense of the senior ally. But because the support thus given can broaden the range of options open to allies and clients, the net effects of this mode of suasion can also be negative. For example, while the U.S. Sixth Fleet may be continuously deterring Soviet and Arab moves against American interests as well as reinforcing the

Alliance, it may also be giving unintended encouragement to Israeli activism in a manner inimical to the interests of the United States.

Given its inherent indirection, the latent suasion of naval forces can therefore produce undesirable side-effects, from the deploying party's point of view. Ground and air forces can also do so, but in their case the latent effects they generate will be as static and geographically limited as the forces themselves. Naval forces, however, because of their unimpeded mobility, entail the possibility of diffused and unrecognized latent effects, including undesirable ones. By the same token, however, once the undesirable effects are recognized, adjustments can be made quickly and silently. If, for example, a routine fleet movement off the coast of India is thought capable of evoking undesirable reactions from the Indian government, the transit route can be shifted to the south, and with none of the rigidities, either technical or political, associated with the redeployment of land-based forces. Thus, the same quality of sea power that is the source of possible political difficulties can also provide, through its inherent flexibility, the means to avoid them, if, that is, any negative political repercussions are in fact perceived. This suggests that continuous political guidance of the highest possible quality is a crucial requirement of overseas naval deployments: a modern oceanic fleet needs a political "radar" as much as it needs the electronic variety.

Even if high-grade political advisers are active in naval commands, and even if they are acute observers of the political "radiation" emitted by the fleet, the problem cannot be fully solved since there may be, and often are, severe distortions in others' perceptions of the fleet—of its tactical configuration and the underlying political intent of its movements.

Generally, political leaders around the world understand more about ground power than air power, and more about the latter than about naval power. In the case of active suasion, decision makers on the other side will at least be faced with a definite grouping of capabilities in the naval forces that are displayed to them, or positioned against them, and the intent of the deploying party will normally be conveyed to them in one form or another. But in the case of latent suasion, neither of these conditions obtains; the political leaders of the littoral states must therefore construe the capabilities and intent of the naval forces which they observe according to their wits, and the possibilities of distortion are vast. For one thing, leaders of the smaller littoral states have ready access to naval expertise only in their own small navies, whose officers may know little about the operation of Great Power navies. With respect to political questions, one national leader is as good as another since each must make his own judgments in the light of his own world view. But assessments of naval capabilities, of the significance of particular tactical configurations, and of the nature of the possible threats emanating from the sea require technical knowledge that many smaller states do not have.

Even those, like the Israelis, who have shown some competence in the conduct and analysis of land and air warfare may have no such expertise with respect to naval matters. There is no land or air counterpart to the naval errors made by the Israelis: the imprudent conduct of the *Elath*, the loss of the *Dakar*, the explosion of ill-loaded landing craft in harbor and others.[5] Moreover, in the many backward countries where military men have seized power, very few sailors seem to have reached positions of political significance, barring the odd Turkish admiral. How do these leaders assess the relative capabilities of the American and Soviet fleets? How do they evaluate the balance of sea power in their region of the world? Do they count ships and ship-classes by referring to *Jane's*? Do they use aggregate tonnage figures, once elevated to the dignity of a treaty standard, but now so obviously misleading in the presence of rapid technological change? In this context, it suffices to say that the relationship between the forces deployed and the suasion effects that these forces actually evoke is neither direct nor proportional.

Since latent suasion is undirected and not overtly linked to any specific policy objective, its net effects, and even more, its net benefits, are difficult to identify and impossible to measure. This would not have disturbed Athenians or nineteenth-century Britons, but given the prevailing tendency to attempt the quantification of the benefits generated by military deployments, the elusive nature of latent suasion is something of a policy problem in itself. In this study, particular attention will therefore be given to the identification and evaluation of the deterrent and supportive effects of latent suasion on adversaries, allies, and third parties.

Active Suasion

Any system of definitions imposed on the fluid and variegated realm of politics will inevitably be both arbitrary and incomplete. But a typology such as the one presented here is only meant to be convenient for analytical purposes: one need not distort every marginal phenomenon in order to obtain a "good fit." With this thought in mind, the exercise of "active" suasion is defined as any deliberate attempt to evoke a specific reaction on the part of others, whether allies, enemies, or neutrals; the reaction actually obtained would constitute the suasion process itself, in this case labeled as active. For example, an attempt to deter an attacker by deploying retaliatory forces and issuing appropriate warnings would constitute the exercise of active suasion, while any deterrence in fact achieved would define the suasion effect itself. This to emphasize yet again that we are dealing here with others' reactions, and not with any objective that can be achieved by the direct application of force in itself.

A very important and much studied form of active suasion, all of it armed but so far only partly naval, is the strategic-nuclear level of deterrence. All forms of deterrence, from that which ensures good manners when foreign warships meet at sea, to that which is intended to avert the lethal threat of strategic-nuclear attack by posing an equally lethal retaliatory threat, are based on the same set of interactions, in which psychological elements are of critical importance: first, the deploying party's prior perception of a threat; second, the development and continuous maintenance of strike-back forces capable, even after absorbing an attack, of inflicting damage deemed to be unacceptable by the other party; third, the target party's recognition of the causal link between the retaliatory threat and the specific actions of his that the deploying party is seeking to avert; and, fourth, the target party's concurrence with the value judgments and technical assessments made by the deploying party. Both must believe that the retaliatory threat is capable of destroying *greater* values than those which the target party can obtain by making the move that is to be deterred.

In the strategic-nuclear arena, these requirements of deterrence have been the subject of so much scrutiny and debate that their reiteration is unnecessary. But this is not so in respect of lower-intensity levels of deterrence: substrategic forms of deterrence are all too often seen in narrow technical terms. While due attention is given to the mechanics of deterrent deployments, the very demanding psychological requirements of deterrence tend to be discounted. *And it is precisely in respect of the latter that deterrence is apt to fail.* If, to quote Mr. Cable,[6] the Israeli destroyer *Elath* was patrolling the coasts of Sinai in 1967 with an "imprudence scarcely rivalled since the ill-fated patrol of the Broad Fourteens by H.M. ships *Aboukir*, *Cressy* and *Hogue*," this was so because the Israelis were convinced that a deterrent was at work to protect their otherwise very vulnerable destroyer. They must have known that Styx-launching patrol boats were in the vicinity; that the *Elath* was sailing well within range of those boats as it turned back off Port Said; and, that the *Elath* was entirely incapable of intercepting or deflecting antiship missiles. But the Israelis tacitly assumed that their own ability to shell the cities and industries in the Canal area rendered their own destroyer immune from attack, as in the case of most other Israeli maritime assets, whose protection has always been ensured by deterrence and not by defense.

All the physical requirements of deterrence were fulfilled, because Israeli artillery was in place and within range. The relative-value requirement was also fulfilled, because whatever the scale of values held by the Egyptian leaders, it was obvious that they could hardly judge the worth of an old Z-class destroyer (symbolism included) as greater than that of the vast Egyptian assets that were hostage to Israeli guns. But the third requirement, recognition of the deterrent linkage, was not fulfilled, and the Egyptians proceeded to sink the ship, only to be surprised by the Israeli shelling of Suez that followed.[7]

Since this was an instance of faulty communication that could have been put right by appropriate diplomatic "signaling" before the event, it is not as instructive as another failure of deterrence that occurred between the same antagonists. Following the sinking of the *Elath* on October 21, 1967, there was almost a year of relative quiet on the cease-fire line along the Canal. The small Israeli troop contingents built no "hard" fortifications; shallow earthworks were deemed sufficient since the line was to be held but not defended against major attacks. More than 900,000 Egyptians lived in the Port Said, Ismailiya, and Suez governorates, the most modern and industrialized of all Egyptian provinces.

To the Israelis it seemed apparent that no conceivable tactical hopes could possibly induce the Egyptians to sacrifice their cities and industries by making the Canal the scene of large-scale fighting.[8] A static offensive by fire was in fact the only option open to the Egyptian leadership, since it was obvious that a Canal-crossing offensive could not be sustained in the presence of Israel's unchallenged control of the air. Nevertheless, the Israelis discounted the probability of an artillery offensive since, according to *their own* scale of values, its limited, almost symbolic, gains would be outweighed by the destruction of the three governorates' cities and industry. But the Egyptians failed to concur in the relative-value judgment made by the Israelis (condition number 4 of deterrence). In October 1968 the Israeli attempt at protection-by-deterrence collapsed when the Egyptians laid down a series of artillery barrages, and sacrificed the Canal-side cities in order to do so. After this bout of shelling, the Israelis were forced to defend the area and to undertake the large-scale construction of "hard" fortifications—at greater cost than would have been the case earlier.

The ultimate source of this error of judgment was the neglect of the significance of cultural differences, and of the differences in relative-value judgments that they imply. Even if it proves to be true that in the strategic-nuclear arena the import of cultural differences is nullified by the world-destroying power of thermonuclear warheads—as one must hope under current offense-only strategic postures—this is certainly not true at lower levels of deterrence. For this reason, and because naval deployments must defend as well as deter in each (substrategic) instance, the body of ideas that has evolved around the theme of strategic-nuclear deterrence is not always a reliable guide to the problems of its somewhat less lethal naval counterparts. For one thing, as it has been suggested, explicit attention must be given to asymmetries in relative-value judgments: the value that our retaliation threatens may seem greater to *us* than the benefit an enemy can obtain by making the attack we seek to deter, but the other side may disagree. This is particularly important in areas such as the Mediterranean region, where two variants of one civilization face an Arab-Islamic culture which is not only very different but which is also undergoing a process of disintegration under the pressure of economic change and nationalism.

This is not the place to discuss the fundamental difference between Western patriotism and Western nationalism as adopted, and adapted, by Third World nationalists. What is highly relevant, however, is the direct implication that this difference between patriotism and nationalism has for a navy that, among its other goals, is seeking to deter. In essence, this boils down to the observed and amply documented fact that for the convinced nationalist, and hence for those who cater to his values as many Third World regimes do, the tangible values of life and property are less important than such intangibles as a valiant self-image and national (for the leader, personal) honor and glory. As many Europeans did until quite recently (and may do again), convinced nationalists discount the material for the sake of immaterial, and the latter often comes in the shape of activist policy goals. Once such goals are declared and duly publicized, they commonly acquire an import that is far greater than the commitments made in the more sober context of Western policy-formation. Moreover, the highly personalized conduct of policy in many Third World countries implies the absence, or at least weakness, of internal institutional restraints such as those which limit the scope of policies elsewhere. Considering recent American experience in such matters, it would hardly do to claim that there is a qualitative difference, but a very significant difference of degree is certainly apparent; here it suffices to underline the cautionary lessons that apply to the conduct of naval deterrence at the substrategic level: first, the requirement of an explicit (but not, of course, public) linkage between the value to be protected and a retaliatory threat; and, second, the desirability of a maximum "value-spread" between the worth of what is to be protected and the adversary values which are threatened in order to deter. This provides insurance against the effect of poorly understood, and always unpredictable, cultural differences.

A second aspect of naval, substrategic deterrence is the duality of mission requirements: if deterrence fails, defense must take its place. In the event of a failure of deterrence, surviving naval forces are not only expected to conduct retaliatory attacks (which may not be launched at all), but also to remain available for subsequent tasks. Instead of the fixed-intensity retaliation that exhausts the mission of strategic-nuclear forces, naval deterrence forces are expected to retain their tactical flexibility. *And this can be applied to the strengthening or even restoration of deterrence, as well as to defense.* It is for this reason that the symbolic use of force must be retained in the arsenal of political rather than war-fighting instrumentalities. If, for example, preclusive control over a given sea area is asserted by threatening to sink any intruding ships, such deterrence can by reinforced by the deliberate display of the forces deployed for the purpose; should this show signs of failing, deterrence can be rendered yet more intense by demonstrating destructive capabilities, e.g., by shooting "over the bows" of any intruders. Finally, deterrence can be salvaged even after a failure by sinking a single ship, in order to deter the intrusion of the rest, and then by sinking a second and a third and so on, until the line is crossed and the objective becomes destruction rather than suasion in any form. Though "limited exchange scenarios," which envisaged the use of thermonuclear detonations for "signalling" purposes, gained some currency in the strategic debates of the 1960s, such ideas are now recognized as mere fanciful speculation. A similar flexibility is, however, a legitimate and useful attribute of (nonnuclear) naval power applied to the exercise of deterrence.

Though much has been made of the difference between "compellence" and "deterrence," they are generally analogous. Both belong to the realm of coercive suasion, a term that underscores the use of the direct threat, and which suggests an affinity with coercive diplomacy.[9]

Both modes of coercive suasion—the negative, or deterrence, and the positive, or compellence—are subject to the same technical and perceptive requirements; both may be tacit or overt; both may be associated with either private or public warnings; and both are subject to the same psycho-political uncertainties. Since the term was first coined by the banks of the Charles River, something of a literature has grown around the concept of "compellence."[10] Its import boils down to the assertion that it is more difficult to compel than to deter since (1) moves are more difficult to reverse than prevent in that the moves to be stopped or reversed may have acquired their own "tactical" and political momentum and (2) public compliance with others' demands would entail additional losses in the way of prestige.

The first purported difference is in fact trivial because, from the point of view of the *offending* party, the factors that lead to a decision to act may be just as forceful as those that may subsequently be brought to bear against a reversal.[11] Tactical momentum may indeed occur in many cases, but *political* momentum is largely conditioned by whether the second difference obtains or not. And it is with respect to the latter that the distinction between deterrence and compellence as forms of coercion seems least warranted, since even if compellence requires actual backtracking rather than the halting of an ongoing activity, it is by no means inevitable that it be made public or be publicly acknowledged. In the case of the Cuba missile crisis, for example, the Soviet leadership was able to withhold any admission of compliance from the Soviet media—and this in the context of the most overt and public exercise of compellence imaginable. More typical is the case of Indochina, where forces may be advanced or withdrawn by an adversary without any overt recognition, public announcement, or indeed the admission that any forces had moved beyond national borders in the first place. *It is apparent that overt, publicly announced deterrence may be more difficult for the target party to comply with than covert compellence, complied with covertly.*

If it is accepted that both deterrence and compellence are modes of the same coercive form of suasion, it follows that the rules of (substrategic) deterrence apply to both. One rule should be mentioned at this point: the survivability of the retaliatory forces must be ensured, but not necessarily in the case of enemy attacks that would entail a much higher, qualitatively different level of conflict intensity. While the strike-back requirement is self-explanatory, the qualification appended here derives from a peculiarity of the age: five[12] nations deploy nuclear weapons, but their use is increasingly deemed, by them, to be unacceptable in all situations other than those which entail a threat to national survival. As a result, these weapons can be excluded from the force-level comparisons made in the context of less-than-vital confrontations, even in a direct and central crisis situation. Thus in order to implement the Cuba quarantine of October 1962, the U.S. Navy did require a clear preponderance over Soviet naval forces in the area, but it could also be assumed that American warships need not defend against nuclear weapons whether deployed *in situ* or anywhere else. Had the crisis resulted in nuclear war, the ships, or some of them, could have been the target of nuclear attacks, but until that point, nuclear weapons of any sort had no direct effect on the naval dimension of the confrontation.

While the fine gradations imagined in the "escalation" scenarios of the past with their complex hierarchy of "thresholds" never did have much validity,[13] it is apparent that in any low-intensity naval confrontation nuclear weapons embodied in the forces deployed on either side will not normally play a significant role in the assessments of the *local* balance of military strength which, together with the balance of perceived interests on either side, will generally determine the outcome of the confrontation. And this will remain so as the intensity of the confrontation increases, until the level is reached where *core* interests come into question.[14] Certainly, as between the super powers, it is only when either or both sees the interests in dispute as reaching the life-or-death level that the strategic-nuclear balance enters into play, and does so decisively.[15] Of this the Berlin crises of the late 1950s and early 1960s are the best examples: the West prevailed, though much inferior in terms of local forces, and though not superior in *perceived* strategic-nuclear terms, because the equation successfully made between the survival of Americans at home and the political freedom of Berlin injected strategic-nuclear power into the confrontation, thus nullifying the significance of the *local* (nonnuclear) superiority of the Soviet Union.

Also performed continuously in a latent form, the active variety of *supportive* suasion is exemplified by the visit of the U.S. battleship *Missouri* to Instanbul in March 1946, which inaugurated the postwar deployment of American naval power in the Mediterranean. This spectacular example of naval symbolism represented an attempt to exercise *active* suasion in a *supportive* mode on the Turkish government, then under severe Soviet diplomatic pressure.

The *Missouri* episode raises one of the fundamental problems of the political application of naval power: the use of ships as symbols, rather than instruments, of power. There is, in effect, a potential contradiction between the well-understood importance of genuine military superiorities *in situ* in determining the outcome of confrontations and the accepted significance of the symbolic warship. How can the symbolic ship that asserts no local military superiority, and whose capabilities may not even be relevant to the setting, secure the interests of the deploying party? What does the ship-as-symbol really symbolize in fact?

Prevailing views on this score are heavily influenced by a received interpretation of the history of British naval supremacy in the nineteenth and early twentieth centuries. This is an interpretation that explains the use of symbolic ships, and much else, purely in terms of naval power: so long as the Royal Navy had a globally superior fleet in the North Sea or anywhere else, a single frigate could effectively impose the will of H.M. Government on recalcitrant coastal states the world over, since the flag it flew was the portent of potentially overwhelming naval force. And that was force that could always be brought to bear if the symbolic frigate was denied its due. According to this view, therefore, the symbolic ship derived its powers from the combat capabilities of the Royal Navy's battle fleet, and its symbolic power was proportional to genuine naval power. This conventional interpretation of the British experience of naval supremacy has now challenged in a most authoritative manner by Gerald S. Graham in his lectures on the "Politics of Naval Supremacy."[16]

Graham's argument is that Great Britain's freedom of action overseas was derived from the paralysis of her European rivals. By virtue of a most successful *continental* policy, the British contrived to neutralize those powers, especially France, which could otherwise have competed with Britain overseas. As a result, the targets of British naval pressures were cut off from all aid on the part of Britain's rivals in Europe, and it was this insulation, achieved by political means, that forced coastal powers overseas to acquiesce in British demands made by a symbolic frigate. Where no such insulation obtained, as in the case of Mehemet Ali's Egypt in the 1830s, the British were in fact unable to enforce compliance with their demands by such means. This implies that the symbolic ship symbolizes national rather than naval power as such; its effectiveness is thus proportional to the former, not to the latter. Naval power is of course a constituent of national power but it need not be the *salient* source of national power, and hence will not define its limits. Nor will it define the power embodied in the symbolic ship. A navally inferior imperial Germany could project her national power by means of a symbolic warship at Agadir; on the other hand, the continuing supremacy of the Royal Navy after the 1870s did not suffice to prevent a decline in Britain's national power, and thus in the impact of her symbolic ships.

Retrospective assessments of the *Missouri* episode, its perceived significance, and its net effects are consistent with this interpretation of the role of the symbolic ship. For all the majesty of her sixteen-inch guns, her enormous bulk, and uniquely strong armor, the specific tactical capabilities of the *Missouri,* or indeed naval capabilities of any kind, were of doubtful relevance to the Russo-Turkish crisis in concrete military terms. The Soviet threat to Turkey emanated from the large Soviet ground forces and from the associated tactical air power that Stalin could deploy on the Turkish-Bulgarian and Russo-Turkish borders. This was the implied threat behind Soviet demands for a renegotiation of the Montreux Convention, and once Turkey recognized that its secular protectors, the British, were no longer up to the task, American support was sought. It is obvious that naval forces could not be injected into this confrontation in a manner that could have altered an exceedingly unbalanced relation of forces between the two sides, since even carrier-based aviation could only have been effective on one half of one sector (the Mediterranean coast of Thrace). But this hardly mattered: the arrival on the scene of the *Missouri* bearing the body of the deceased Turkish ambassador to Washington, was not intended to alter the local balance of power but to affirm a commitment. The *Missouri* was an American warship and its journey was a splendid diplomatic gesture that enabled President Truman to make a commitment without "benefit of Congress."[17]

Less sophisticated observers than the Turks or the Soviets may have been impressed by the sheer size and formidable guns of the *Missouri,* but in this case the warship symbolized presidential willingness to take over the traditional British role in protecting the Straits' status quo, and this sufficed to establish an American commitment as far as the Turks were concerned. (Until then Turkish-American relations had been poor, the main subject of diplomatic communications having been the insistent and futile wartime demands made on the Turks to suspend mineral ore shipments to the Germans. After the *Missouri* episode, however, the Turks felt free to reject Soviet demands, acting on the assumption that they had found a new protector in the West.)

We do not know what (coercive/deterrent) impression the *Missouri's* visit made on Stalin's government, but we do know that Soviet diplomatic and propaganda pressure on Ankara continued and went on doing so even after the task force built around the aircraft carrier *Franklin Delano Roosevelt* arrived on the scene, six months after the journey of the *Missouri.* On the other hand, by the end of 1946, before the formal enunciation of the Truman Doctrine and its congressional endorsement, the Soviet political offensive against Turkey petered out.

All this suggests that the symbolic warship can play its role only before, and in order to prevent, a confrontation. Its effect on the *local* balance of power may be insignificant, but its purpose is to affirm a commitment of national power, local and strategic, naval or otherwise. In this instance, the symbolic warship was used to affirm a commitment promptly and tangibly in circumstances when political conditions in the United States were such that President Truman was not inclined to seek congressional approval for a formal commitment by treaty of alliance or unilateral declaration.

If such commitment making fails to avert a confrontation by its coercive/deterrent suasion effect, then the contest would undergo a change, and only genuine local superiorities would matter in determining the outcome (subject to the rule-setting implications of the strategic balance). It is apparent that nonexistent commitments cannot be simulated by the skillful manipulation of symbolic warships or artful diplomatic image making. Had the Soviets felt in 1946 that the maintenance of the status quo in Turkey and the Straits was not considered an important American national interest by the White House, and prospectively, on Capitol Hill, the *Missouri* visit could have had no impact on their own decisions with respect to Turkey. (Perhaps this is what they *did* believe, since Soviet pressure on Turkey did continue.)

Making a commitment implies the willingness to resort to force, and if the setting calls for it, large-scale force. Without a preponderance of deployable national power and the intent to use it, gestures intended to affirm commitments will fail. Thus a trip by the *Missouri,* or indeed a whole task force, to Seoul in 1949 would not have averted the North Korean invasion (nor would a different text in Acheson's National Press Club speech) had the Soviets and North Koreans felt that South Korean independence was not in fact a salient interest in American eyes. In a marginal case, as South Korea perhaps was, gestures could have been of use to signal intent, but only if they were congruent with policy (and domestic) attitudes—as perceived by others. In this respect, American willingness to accept the inevitable in China seemed to indicate a similar disposition with respect to Korea (provided that a communist victory was made to seem equally inevitable); this was the "signal" that was perceived on the other side.

The inherent danger that supportive suasion may encourage the supported party to go too far has been discussed above and it was concluded that a restraining hand was often a necessary adjunct to the valiant arm. With respect to the active form of supportive suasion this same danger is easier to recognize.

In the 1958 Middle East crisis, Soviet policy faced precisely this problem. To avoid an unwanted intensification of Egyptian activism, then already manifest in the attempt to subvert Lebanon and Jordan, Khrushchev reportedly told the Egyptians that the Soviet effort to dissuade Anglo-American intervention against revolutionary Iraq, and in the Levant, would not extend to the use of force on behalf of Egypt.[18] Once a great power acquires a client in a region, any attempt to apply deterrent coercion on other small powers in the area will evoke supportive effects upon the client, if both the great power and the client face the same adversaries. This is often a positive factor insofar as it strengthens the alliance relationship, but the client may also go too far from the senior partner's point of view; the client may try to exploit the patron in the expectation that forces will be made available to the client too, to protect it from imminent destruction.

By making it clear that his threats were strictly verbal, Khrushchev was trying to negate the supportive suasion inherent in his attempted exercise of coercive suasion against the United States and Britain. His warning that Soviet help would not be forthcoming in the event of an Anglo-American attack on Egypt was calculated to dissuade "adventurism" in Cairo, and it was apparently reasonably successful. Had the Soviets had a fleet on the scene, they could have tacitly achieved the same effect, but forcefully, by ordering Soviet warships out of Egyptian ports and as far away as possible (so long as this move was not inconsistent with the attempt to deter Anglo-American intervention).

In 1950, elements of the U.S. Seventh Fleet were deployed between Taiwan and China proper to dissuade, and if need be, defeat any attempt at an amphibious invasion of Taiwan. At the same time, unintended supportive effects upon the Nationalists were neutralized by the public declaration that the Fleet would also intervene to prevent a Nationalist landing on the mainland. In this case, the deterrent effect on the Peking government was actually reinforced by the careful delimitation of the supportive role of the Seventh Fleet: its incentive to attempt an invasion of Taiwan was much reduced once the Nationalist threat to the mainland was neutralized, and a "virtuous circle" was thus set in train. Here again, the peculiar flexibility of naval forces enabled them to play a precisely defined role in a manner that has no equivalent on land.

In the context of the 1969–70 Egyptian-Israeli War of Attrition, for example, when the Soviets moved their antiaircraft missile regiments to Egypt to dissuade Israeli activism, they were automatically, and perhaps unintentionally, encouraging Egyptian activism. When the deployment was in place, complete with the rudiments of an integrated air defense system, fighter squadrons and the Goa batteries, the Soviet forces were there for the Egyptian leadership to use, since these forces would have been forced to intervene in the event of a conflict, if only to defend themselves. And this was so whether they faced an Israeli air offensive (the move to be deterred) or a counter-offensive precipitated by an Egyptian artillery offensive, which the Soviet deployment rendered more probable, whether the Soviets wanted such an offensive or not.

As we have seen, the duality of effects produced by active suasion in a coercive mode presents a typical problem of alliance statecraft in the almost inevitable supportive effects thereby evoked. And the theoretically defensive nature of particular deployments is no insurance against their manipulation by junior partners. In the case of naval forces, moreover, the purported distinction between defensive and offensive forces lacks even a semblance of validity. At the same time, however, the flexibility of naval forces enables the deploying party to make precise mission distinctions. Moreover, should that not suffice, the senior partner can always order its fleet out of the area, thereby disabusing a client who wants to misuse a deployment meant only to deter others. And this can be done within a much shorter period of time and with none of the political costs or tactical dangers entailed by the withdrawal of land-based forces.

Since 1945, the U.S. Navy has exercised active suasion, support or coercive, deterrent or compellent, on more than seventy occasions[19] at all levels of intensity and upon areas of the globe ranging from the Caribbean to North Korea through Trieste. In each case the *sine qua non* for the successful evocation of the desired response has been the other side's perception of its warships, their tactical configuration and their intent.

Notes

1. As defined in the 1972 and 1973 posture statements; e.g., Statement of Admiral Elmo R. Zumwalt, Jr., USN, Chief of Naval Operations, on Department of Defense Appropriations, FY 1973, U.S. Congress, Senate, *Hearings before the Senate Committee on Appropriations,* 92nd Cong., 2d Sess., 1972, Pt. 3-Navy, p. 67.
2. There is no shortage of such terms. Even "presence" has an unfortunate connotation in that it implies physical visibility where none may exist. More important, it suggests passivity where none may be intended—or perceived. One typical erroneous deduction is that submarines are inherently unsuitable for "presence" missions. In fact, even a ballistic-missile submarine has been used to assert a (strategic-nuclear) "presence." (For example, the SSBN "Patrick Henry" visited Izmir in 1963 to reaffirm the inclusion of Turkey within the scope of strategic-nuclear deterrence following the removal of the land-based IRBMs previously deployed on her soil.) The literature also includes "interposition," valid enough to describe instances of effective blockade but otherwise anachronistic in its associations with the age of line-of-sight gunnery. A British battleship could no

longer rely on "interposition" to protect (Basque refugee) ships from a modern counterpart to the nationalist cruisers of the Spanish Civil War, if the latter were equipped with maneuverable SSMs—unless that is, "interposition" implies a direct (deterrent) threat which would in fact deprive the concept of its peculiar meaning, and analytical worth. A new classification is presented in James Cable's recent study, *Gunboat Diplomacy: Political Applications of Limited Naval Force* (New York: Praeger for The Institute for Strategic Studies, 1970). His definitions, "definitive," "purposeful," "catalytic," and "expressive" force, intermingle functional and intensity criteria. As a result they are more useful for descriptive than for analytical purposes.

3. Vice Admiral Stansfield Turner, President (1974) of the U.S. Naval War College, suggested this point in conversation.

4. This "resolve" to use force is not a generalized psychological propensity but rather the reflection of a given policy priority. This priority in turn must derive from an assessment of the importance of the object of a confrontation. The *credibility* of a party in the eyes of others, i.e., his resolve, will depend on others' estimates of how he views the interests in dispute. One cannot, therefore, hope to "show resolve" or augment one's credibility in a confrontation by the artful manipulation of images and the use of signals conveyed by means of symbolic force independently of the importance of the *object* of a confrontation. There is, of course, some scope for the diplomatic manipulation of others' policy images so as to magnify one's interests in others' eyes, but the rationality nexus cannot be cut entirely. Thus, the defense of Berlin can be assimilated to the survival of the United States by prolonged and successful diplomatic image-making and the old dramatic gesture (Kennedy's "I am a Berliner"), but few localities in dispute can be turned into a Berlin which strategic-nuclear forces alone suffice to protect.

5. In the 1956 Sinai Campaign a key objective, Rafah, was to be attacked on land in the wake of naval bombardment by a French destroyer. The Israeli command assumed that the bombardment would annihilate the Egyptian defenses even though a single destroyer (three twin five-inch guns) was involved. One hundred and fifty shells were fired, and with little effect. In his *Diary of the Sinai Campaign* (London: Weidenfeld and Nicolson, 1967), p. 74, Moshe Dayan candidly admitted that his expectation had been based on memories of the war films that he had seen, and little else.

6. Cable, *Gunboat Diplomacy*, p. 76.

7. It was the contemporary assessment of the Israelis that the Egyptians had been surprised by the *nature* (and intensity) of their response. Comments in the Egyptian press at the time support this deduction. See Muhamad Hasanain Haykal in *Al Ahram*, 23, 24 October 1967.

8. See Edward N. Luttwak and Dan Horowitz, *The Israeli Army* (New York: Harper & Row, 1975), for a narrative of this episode.

9. In fact, the two concepts relate to different aspects of the same political phenomenon; each covers one half of a whole that is separated only by institutional boundaries.

10. The term "compellence" was first used in this sense by Thomas C. Schelling in his *Strategy of Conflict* (Cambridge, Mass.: Harvard University Press, 1960). The term has since acquired general currency.

11. This can be taken further. Where national leaders have to contend with a "street" opinion that is both activist and ill-informed, the only way of dealing with activist pressures may be to act, and then use of others' threats—or actions—as evidence that it is the force of circumstances, and not personal corruption or cowardice, that prevents a more active policy on the part of the national leader.
12. Indian plutonium fission devices are not to be counted as weapons—or so the Indian government says.
13. Since they ignored the effect of perceptual differences between the two sides, among other things. Here it has been argued that across cultural barriers men may see elephants where there are only mice and vice versa, while the "threshold" concept assumed that men would respond to obscure differences between breeds of mice—or elephants.
14. And then the determining factor will be the perceived *strategic*-nuclear balance—as well as the balance of perceived interests.
15. For a brief discussion of the issue, see Edward N. Luttwak, *The Strategic Balance 1972* (New York: The Library Press for The Center for Strategic and International Studies, Georgetown University, 1972), pp. 69-92.
16. Gerald S. Graham, *The Politics of Naval Supremacy* (Cambridge: Cambridge University Press, 1967).
17. Ferenc A. Váli, *The Turkish Straits and NATO* (Stanford, Calif.: Hoover Institution, 1972), pp. 58-81: and Feridun Cemal Erkin, *Les Rélations Turco-Soviétiques et la question des détroits* (Ankara: 1968), pp. 286-373.
18. Muhamad Hasanain Haykal, *Al Ahram,* 22 January 1965.
19. A partial list, which stops short at 1969, and may otherwise be incomplete, is found in the addendum of a letter from Admiral T. H. Moorer to Senator Mondale, reproduced in U.S. Congress, *CVAN-70, Joint Hearings before the Senate and House Armed Services Committees,* 91st Cong., 2d Sess., 1970, pp. 163-65.

7

Sea Power in the Mediterranean

The Rules of the Game: The Central Balance and Its Implications

Until the mid-1960s, all the warships in the Mediterranean were either American, allied or else insignificant. Accordingly, the supremacy of the Sixth Fleet was seemingly uncontested. But in fact the Soviet Union already had a large submarine fleet, and also many land-based naval strike aircraft. In spite of both technical and geographic limitations, these submarines and aircraft amounted to a serious military threat for the Sixth Fleet. Why then was this threat ignored, or at least heavily discounted? After all, the deployment of Soviet W-class submarines and Tu-16 naval bombers within striking range of the carrier task-forces in the eastern Mediterranean was no secret. Only the most optimistic observers could assume that no Soviet submarines would be on the right side of the Turkish straits on the first day of the war, or that all Tu-16s would be intercepted in good time.

Whether by systematic reasoning or intuitive understanding the political observers whose views in fact determined the perceived power of the Sixth Fleet (and therefore its real political weight), discounted Soviet anti-shipping capabilities because they saw a drastic political constraint operating upon them, a constraint much more severe than any tactical or technical shortcoming: W-class boats and Tu-16s could not be employed by a rational Soviet leadership against the Sixth Fleet except in the most extreme of circumstances.

For all practical purposes, there were only two plausible scenarios in which a direct attack upon American naval forces was rational from the Soviet point of view: either in the context of a general attack upon American strategic and forward-deployed forces, or, alternatively, in the event of an attack by these same forces against the Soviet Union. The first of these scenarios, while rational in principle, was exceedingly improbable given the strategic-nuclear inferiority of the Soviet Union in those days, especially in a counter-force exchange. The

101

second scenario was inherently improbable and, moreover, devoid of military significance: once the navy's A-3s or A-5s bearing nuclear weapons had been launched from the deck of the carriers, the strategic mission of the Sixth Fleet was exhausted.

A military force that could only be used in such extreme and unlikely situations was also a force whose political utility was highly restricted. The Sixth Fleet could control the sea lanes, it could project air power ashore, it could land troops and it could shell coastal targets. Soviet Tu-16s and W-class boats could do none of these things; neither could they prevent the Sixth Fleet from exercising its varied capabilities, for an attack upon the fleet was only conceivable in the context of the extreme scenarios of strategic attack, and strategic (would-be) preemption. The Sixth Fleet, on the other hand, had a full range of usable tactical options against third parties, and it could therefore be used as a highly flexible instrument of armed suasion. Soviet naval forces by contrast only had the all-or-nothing option of strategic attack against the forces of the United States; and they had only anti-shipping capabilities against third parties, or a limited coastal-defense capability on their behalf.

There were thus two dimensions to American naval superiority in the political arena: first, the flexibility of the forces, i.e., the variety of their tactical options, as opposed to Soviet inflexibility, and, second, the validity of American naval power. As a valid instrument it could be used to support or coerce third parties in a variety of ways, while Soviet naval power offered only two tactical options useful to support or coerce third parties: first, coastal defense on behalf of Soviet clients, and second, a submarine anti-ship capability that could be used against the enemies of these clients. Given its definitive nature, and the inherent lack of display potential of submarine forces, this highly restricted option was not very useful in the political arena in the absence of hostilities.

This, then, was the setting in which the Sixth Fleet was seen as a powerful political instrument endowed with important capabilities and with the necessary freedom of action for their use.

With the solitary exception of the 1958 landings in Lebanon, this freedom of action was never actually exploited. But this scarcely describes the political utility of the fleet. In fact it was engaged in political operations almost continuously as the chief tool of armed suasion for American foreign policy in the region. Its latent presence was of permanent importance, while its active suasion played an important role in every one of the many crises in the Mediterranean in the postwar period, as the following partial list shows.

TABLE 1

The Political Uses of United States Naval Power in the Mediterranean, 1945-1975 (a partial list)

TIME	AREA	PRIMARY GOAL	SECONDARY GOAL
1. April 1946	Turkey	Support T's resolve.	Deter USSR.
2. July 1946	Adriatic	Deter Yugoslavia.	Deter Italy.
3. September 1946	Greece	Support Greek Government.	
4. 1946-1949	Greece	Latent support for Greek Government.	Deter USSR.
5. May 1956	E. Med	Support Government of Jordan.	Deter Egypt.
6. Oct.-Nov. 1956	E. Med	Deter Israel, France, U.K.	Deter USSR.
7. April 1957	E. Med	Support Government of Jordan.	Deter Egypt.
8. August 1957	E. Med	Support forces in Syrian politics.	Deter Egypt.
9. May 1958	Lebanon	Support forces in Lebanese politics.	Deter Egypt.
10. August 1958	E. Med	Support Government of Jordan.	Deter USSR.
11. April 1963	E. Med	Support Government of Jordan.	Deter Egypt.
12. June 1967	E. Med	Deter U.S.S.R.	
13. September 1970	E. Med	Coerce Syria; Deter U.S.S.R.	Support Jordan.
14. October 1973	E. Med	Deter U.S.S.R.	

Whether or not this particular categorization is accepted in every detail, it will no doubt be generally agreed that the suasion exercised by the Fleet rarely had the Soviet Union as its primary target. It is also clear that the particular capabilities that made these exercises of naval suasion credible were precisely the capabilities that the Sixth Fleet had and the Soviet Union lacked. It is only in the very first case here listed that the use of the Fleet was largely symbolic. For in the Russo-Turkish crisis of April 1946 the deployment of American naval forces to Turkish ports was essentially an expression of commitment, without benefit of Congress; their actual capabilities hardly mattered.

While such naval suasion was exercised through the political "tactics" of directed deployments, force-level changes, display maneuvers of various kinds, operating changes and operative acts (i.e., intrusive reconnaissance), the credibility of these actions rested ultimately on their relationship to actual capabilities, which were (a) usable, (b) important in the context, (c) not obviously capable of being negated by local (or Soviet) counteraction. For example, in case two, American naval forces could have blockaded Yugoslav ports or simply denied passage to ships bound for Yugoslavia in the straits of Otranto. Gunfire from American warships off Trieste could probably have interdicted any attack on that city, since it could have reached everywhere in the very narrow hinterland of the Italian Zone A. These tactical options would have been of considerable importance to the outcome of an actual conflict and they could not have been negated by the adversaries: American aircraft could have controlled the Adriatic air space from Italian bases and effectively prevented Russo-Yugoslav air attacks upon U.S. naval forces while Yugoslavia had no significant naval forces of its own. Nor could Soviet submarines have survived in the very shallow Adriatic even if they had been successful in reaching that area of operations.

Since the mid-1960s, the circumstances of the Sixth Fleet have seemingly changed drastically. On the one hand, the strategic nuclear balance is now set at an uncertain parity. This presumably means that the extreme strategic attack scenario discussed above has become more realistic from the Soviet point of view. Other things being equal, an all-out Soviet attack against American strategic, and forward-deployed, forces is now theoretically more probable because it is likely to be more successful. But of course other things are not equal. In fact the entire political context makes a would-be disarming Soviet counterforce attack even less likely than before. And this is the only consequence of the change in the strategic balance for the naval balance in the Mediterranean, though admittedly an important one (Luttwak, 1972).

The other change in the balance of forces is altogether more salient in the calculations of those who evaluate the relative strengths of naval power in the region. This is, of course, the growth and forward deployment of Soviet naval forces in the Mediterranean. On the face of it, the inflexibility of Soviet naval power is a thing of the past. Much-strengthened submarine forces and modernized naval-air forces have now been joined by a forward-deployed surface fleet of missile destroyers and cruisers and helicopter carriers. On the face of it, this new Soviet naval strength is such that (a) it can negate American naval tactical options and thus undermine the credibility of American naval suasion, and (b) it can exercise naval suasion as an instrument of Soviet foreign policy (Zumwalt, 1975:861). An explicit and not excessively modest claim on these lines has been advanced by the commander-in-chief of the Soviet Navy.

The assertions of Admiral S. G. Gorshkov have been accepted in toto by many observers; they have even been promoted by a former Chief of Naval Operations. It has become the common wisdom of the day to assert that the arrival of the Soviet Mediterranean fleet has had a crippling effect on the freedom of action of the Sixth Fleet. For example, it has often been said specifically that an act of overt intervention such as the 1958 landing in the Lebanon would no longer be feasible. Presumably by this it is meant that such an action would now entail unacceptable risks. If this were true, it would logically follow that the threat of an American amphibious landing in Lebanon would lack credibility, thus presumably being ineffectual as a tool of armed suasion. There are here three quite distinct questions. First, is it true that in a conflict the Soviet naval forces which could be brought to bear in the Mediterranean would incapacitate the Sixth Fleet? Second, if this is true, does such a capability ipso facto negate the suasion potential of the Sixth Fleet? Finally, as a separate question, does the Soviet Fleet have a suasion potential of its own, and if so of what kind?

Much of the confusion now prevalent in discussions of the political utility of the Sixth Fleet arises from the failure to make the necessary distinctions between these three questions. For the first question, the ultimate military question, no generic and clear answer can be given. Much depends on the exact composition of the Soviet and American forces deployed in the Mediterranean at the time of the conflict. In October 1973, it was reported that there were 23 Soviet attack submarines, 29 surface combatants, and 44 other vessels in the Mediterranean (Moorer, 1975:271). This is a vast array, and the large Soviet naval air force could have added a considerable long-range anti-shipping capability. Moreover, the deployment of short-range Soviet fighter-bombers in Syria and Egypt could not then have been excluded.

Would such a combined force have sufficed to incapacitate the Sixth Fleet in combat? Much would depend on the exact sequence of events. It is obvious that with its non-reloadable anti-shipping missile launchers, the Soviet surface fleet can only expect to fight successfully in a surprise attack scenario, i.e., the "splendid" all-out missile strike stressed in worst-case analyses of the naval balance. It is equally obvious that the carriers of the Sixth Fleet could achieve the same results without the benefit of surprise.

Fortunately, it is not necessary here to carry out a systematic analysis of the tactical balance of forces taking into account air, surface, and submarine capabilities, varying scenarios, and so on. Such elaborate evaluations are useful exercises for the purpose of naval planning, but they are only tangentially relevant to the question at hand, i.e., the ability of the Soviet Navy to negate the suasion potential of the Sixth Fleet and so nullify its political utility.

Soviet cababilities vis-à-vis the Sixth Fleet do not directly detract from its suasion potential because it is only in particular, extreme, and most unusual circumstances that the Soviet threat to the Sixth Fleet could be activated by a rational Soviet leadership. This is because a partial attack would lead to a crushing Soviet defeat, while an all-out attack overtly made by Soviet forces against American warships would entail—under almost any conceivable circumstances—a very high risk of American retaliation with forward-deployed nuclear forces, or even strategic forces in selective use. A high risk of nuclear retaliation can only be accepted by a rational Soviet leadership if the survival of major Soviet territorial or military assets is in question. Hence, the Soviet Union cannot credibly undermine the suasion potential of the Sixth Fleet by depriving its tactical options of credibility—unless in an extreme crisis, where core Soviet values are in dispute. It follows that in the context of all the regional crises listed above except the first, and indeed in the context of the regional crises that are likely to erupt in the future, the ability of Soviet naval forces to mount a "splendid" surprise all-out attack cannot affect the political utility of the Sixth Fleet.

It may be useful to illustrate this analysis with a miniscenario, a contemporary version of Lebanon 1958, the very operation that is now so widely thought to have become infeasible: assume a restored pro-Western regime in Lebanon, assume a restored Western interest in the status quo, assume an attempt on the part of the Palestine Liberation Organization (PLO) to seize power, assume Syrian support for the PLO, assume a vigilant Israeli neutrality, and, finally, assume a full forward deployment of the Soviet Fleet in the eastern Mediterranean, as in October 1973.

U.S. Moves I:

The Fleet is ordered to deploy two Marine battalion landing teams, and to be ready to provide counter-air (vs. Syria), and close-support for the Marines as needed. The Sixth Fleet approaches the operational area; amphibious shipping moves into place.

Russian Options:

1. Inaction.
2. Direct attack upon the Sixth Fleet, or the landing force per se.
3. "Interposition"? (Physically intercept American ships? Shoot across their bows?) "Interposition" is in fact governed by the same rules as option two, i.e., it can only be operative if the strategic risks to the Soviet Union are warranted. Is the success of the PLO in changing the Lebanese status quo a core Soviet value (i.e., is either the success of the PLO *or* a change in the Lebanese status quo a core interest whose attainment warrants nuclear risks?) Short of reviving the naval tactics of the fifth century B.C., there is in fact no genuine (i.e., nonshooting) form of interposition. Hence, the concept of interposition is void of meaning.

U.S. Moves II:

The Sixth Fleet continues to approach the area. American amphibious shipping moves to the shoreline. The Soviet Fleet adopts an attack configuration.

At this point the Soviet Union has any number of non-naval options, from propaganda and political warfare in general to the introduction of air and ground combat units into Syria to support the Syrian and the PLO forces. But the Soviet Fleet has no definitive options per se. It can only influence the situation if its deployment succeeds in intimidating American decision makers, i.e., if its threat-maneuvers have a psychological impact upon them that outweights in their minds the logic of mutual deterrence. This, it is to be hoped, will not happen.

The landing operations continue as per the 1958 scenario.

This then is the consequence of the *balance of perceived interests.* The values in dispute being much less important than the Soviet values which would be prejudiced by an attack upon the Sixth Fleet, the Soviet Fleet is condemned to inaction, given the balance of strategic-nuclear power and given the magnitude of the provocation that an attack upon the Sixth Fleet would constitute.

All this is not to say that the Soviet Fleet is ipso facto incapable of neutralizing the suasion potential of the Sixth Fleet in any and all conceivable circumstances. Notably, if the Soviet Union were to achieve a conclusive superiority in the strategic-nuclear realm (i.e., if it were to acquire a unilateral advantage in counterforce capabilities against, say, the American land-based missle force), then it could hope to launch an all-out attack upon the Sixth Fleet without necessarily being subject to nuclear retaliation. In other words, if the Soviet Union had dominating options at the strategic-nuclear level it could then expect to deter American escalation to that level of conflict intensity, thus forcing the United States to accept an attack upon the Sixth Fleet as a fait accompli unless other valid options were available to the United States.

The Soviet Union does not yet enjoy any such advantage at the strategic-nuclear level. Moreover, even if it did, an attempt to exploit "escalation dominance" in this manner would still entail very grave risks, for even then not all American options—including nuclear options—would be foreclosed. For example, in an all-NATO crisis, forward-deployed nuclear forces such as the FB-111 based in the United Kingdom could be used to respond to an all-out strike against the Sixth Fleet by attacking Soviet naval installations on the Kola Peninsula. Nevertheless, given an imbalance at the strategic-nuclear level and a corresponding psychological climate (for third-party decision-making especially), there is no doubt that the suasion potential of the Sixth Fleet would be drastically curtailed, for its credibility would then have been undermined. By contrast, in conditions of strategic-nuclear parity, the suasion potential of the Sixth Fleet can only be affected by Soviet counter-suasion if the values in dispute are, or in the course of the crisis become, "core" values, on whose behalf a conflict featuring the use of nuclear weapons becomes conceivable (through the dynamics of commitment making).

The difficulty here is of course the definition of "core" values. Competent observers will agree that in the circumstances of 1979 neither the success of the PLO nor the Lebanese status quo can remotely approach the status of core values for the Soviet Union. It would likewise be generally agreed that the physical integrity of, say, Odessa, is definitely a core value, so that the Soviet Union could negate an exercise in suasion which rested on a threat against Odessa because the threat would not be credible—unless American core values were in dispute. But between such extremes there is a broad spectrum, and much too broad a range of uncertainty over what are, and what are not, core values. Certainly, the line cannot be drawn at the boundaries of homeland territories; for example, in Europe and in the case of Japan, explicit American nuclear guarantees are operative. It is not the guarantees that have made the physical integrity and political independence of such non-homeland territories core values for the United States. Rather, it is their prior status as core values that has made American nuclear guarantees credible.

How does a core value come to be recognized as such? How for example was Berlin promoted from a vulnerable outpost to a core value? It is a matter of record that between 1948 and the present, the status of West Berlin has been that of a core value for the United States. It is an observed feature of Soviet conduct that the Soviets have acted on the assumption that a seizure of the Western sectors of Berlin would entail American nuclear retaliation. As a result, the Soviet Union has not been able to exploit its vast military superiority in the immediate area to exercise any perceptible degree of suasion over German or American (and NATO-wide) decision making over Berlin.

Is it conceivable that the Soviet Union could create a West Berlin of its own, defensible by means of strategic deterrence alone, say, in the Middle East? The answer cannot emerge conclusively from an examination of tangible "national interests" and of their "real" weight. The definition of what is and is not a national interest is always arbitrary; the concept will not support an analytical edifice. Nor can interests be qualified as real or unreal, least of all in terms of their tangible worth. Values, i.e., interests, have no objective existence; they are defined by the expression of commitment, and core values (vital interests) are defined by the expression of inflexible and open-ended commitment.

Interests or values, and core values or vital interests, are defined subjectively, and their status is subjectively accepted or subjectively denied by the other side. For example, the successive leaders of the Soviet Union have reiterated ever since 1948 their total commitment to the preservation of the client-regimes of Eastern Europe, if need be by force. For years, successive American administrations have reiterated an inflexible commitment to the security of Western Europe, of Japan, and of Israel. These expressions of commitment have been made, and in the case of the American commitments they have received widespread public support and repeated congressional endorsement. When such commitments were challenged, repeatedly in the case of Berlin, they were reaffirmed in word and symbolic deed; in addition, concrete steps were taken to enhance the capabilities needed to fulfill the commitment.

By a reverse process, failure to reiterate commitments on requisite occasions—for example, by signalling that policy changes have been made and that this or that commitment is "no longer operative"—core values (= vital national interests) can be demoted to mere values (= national interests) or even allowed to wither altogether. The gain in such cases is the foreclosure of avenues of involvement with their inevitable costs. The loss is the loss of power. For beyond the national territory itself, whose integrity is per se a core value, all else must be secured by acts of commitment, unless it is secured more painfully by an active defense in the case of an actual conflict. The United States has power beyond its shores to the extent that commitments have been made. The inflexibility that results is inherent; "full flexibility" amounts to impotence, for in the absence of recognized commitments nothing at all can be held—unless it is being actively defended, by the direct force of arms.

It is quite apparent that for all the attention and all the costly aid that the Soviet Union has lavished upon Algeria, Egypt, Syria, Iraq, and the North Yemen, none has been elevated into a Berlin in the eyes of either Russians or Americans. At present, Soviet relations with these countries no longer even have the character of a loose alliance, as was once the case, though this could again change and very quickly. But this is not the fundamental point. A Berlin, that is, a core value that ipso facto becomes defensible by the implied threat of nuclear retaliation alone, (because its protection is thought to warrant the risk of nuclear war), has to be created in the minds of potential adversaries by deliberate acts of commitment-creation—acts that potential adversaries eventually come to recognize. So far, the Soviet Union has not engaged in any such effort.

The Soviet Union could not, for example, credibly protect Syria against the Six Fleet by invoking the threat of nuclear retaliation. If units of the Sixth Fleet were to shell Syrian coasts, or, in a less improbable scenario, if carrier-based American aircraft were used to strike at Syrian forces in defense of Lebanon (or Jordan, or Israel), the Soviet Navy could not intervene directly against the Sixth Fleet, for that again would entail risks that would be unacceptable, given the nature of the value in dispute. It is, of course, conceivable that in the future the Soviet Union might seek to make Syrian security a core value, beginning with the signature of a treaty of defense providing for direct military intervention (and not "consultation," as in the present treaty), and continuing through all the diplomatic maneuvers of long-term commitment creation. This has not happened as yet.

It should be noted that aside from the dominating political reality of the "balance of perceived interests," which undermines the credibility of any Soviet attempt at counter-intervention vis-à-vis the Sixth Fleet, tactical factors also militate against the utility of the Soviet Fleet in this respect.

Owning to the specific structure of the two fleets, Soviet naval forces can do virtually nothing against American naval power unless the very first thing done is the destruction of the Sixth Fleet's aircraft carriers, billion-dollar assets each with some five thousand Americans aboard. It is this very substantial tripwire which raises the threshold of any Soviet attack upon the Fleet to the point where the "balance of perceived interests" and the strategic-nuclear balance come into play.

The known structure of the Soviet surface fleet as a one-shot surprise attack force means that it would be extraordinarily difficult to employ in a limited fashion, e.g., to signal firm intent. The American Navy would have to treat any missile firings that were not unambiguously accidental as the ragged prelude to an all-out strike, and would presumably respond by sending attack aircraft to destroy every missile platform within range. In other words, the tactical inflexibility of Soviet Naval forces precludes their use in a confrontation unless the full consequences of an all-out naval war are indeed acceptable to the Soviet leadership. This leaves only the option of submarine attack, overt or covert, i.e., the same options that the Soviet Union had even before the forward-deployment of its surface forces. It is for these reasons that the "active suasion" potential of the Sixth Fleet has not been negated by the Soviet Fleet: threats of counter-intervention designed to undermine its credibility would themselves not be credible.

It has been claimed that its surface fleet has already proved to be more useful to the Soviet Union than these observations would suggest. For example, Soviet spokesmen have said that in the wake of the coup that overthrew the royal regime of Libya on September 1, 1969, Soviet naval forces carried out an "interposition" maneuver to deter American military intervention in Libyan affairs. There is no evidence that any intervention was intended. But again, had there been an intervention, featuring presumably amphibious landings to seize control of Tripoli and Bengasi, how would Soviet naval units have "interposed" themselves? It is after all most unlikely that the Soviet leadership would have accepted the risk of opening fire upon American warships in order to protect a newly established Libyan regime of which it knew nothing. (And which in the event turned out to be fiercely anti-Soviet.)

The reader will have noted a certain tendency to emphatic statement and reiterated explanation. Both are motivated by the prevalence of error as regards these matters. For example, a former very senior officer of the United States Navy has of late chosen to assert that American conduct in the 1973 Arab-Israeli conflict was constrained by a sense of naval inferiority. The latter, in turn, presumably reflected an assessment of the tactical balance of naval forces, as deduced from the projected outcome of naval battles. The logical fallacy is obvious: a calculation of forces based on purely tactical assumptions without reference to political considerations can tell us nothing whatsoever about the political consequences of any tactical balance of power. For example, it would have been quite easy to demonstrate that Western forces in Berlin were inferior to those of the Soviet Union in 1961 as before and since. But the outcome of the Berlin crisis reveals that the tactical balance of power was entirely irrelevant; it was the balance of perceived values that determined the outcome, not the balance of locally deployed forces. Equally, in the course of the 1973 Arab-Israeli conflict, the United States had a value in dispute, Israel's survival, that outweighed any values on the Russian side, and it is this factor which determined the balance of perceived interests.

It is exceedingly unfortunate that misguided interpretations of the meaning of sea power in the region have been encouraged by the deliberate exaggeration of the risks of a "nuclear confrontation" between the superpowers. For complex and not ignoble reasons, official American commentaries on both the 1970 Syrian-Jordanian crisis and the 1973 Arab-Israeli War have stressed the imminence of a direct Russo-American conflict. In spite of these widely circulated official opinions it is the virtually unanimous view of expert opinion that the risk of a direct armed conflict between the superpowers was in fact very small on both occasions because the Soviet Union had no core values in dispute. Analyses by area experts of the "DEFCON Three" subcrisis (Cline, 1974), of the 1970 crisis (Blechman and Kaplan, 1978), and of the dangers of a direct Russo-American conflict over the Middle East in the foreseeable future (The National War College Proceedings, 1975), are united in rejecting the authenticity of the danger. On the other hand, it is perfectly clear that American policymakers saw it as beneficial to intensify the perceived gravity of the crisis, both during its course and also afterwards. Unfortunately, the manipulation of the evidence performed for the occasion has had a lingering effect.

It is, of course, possible that on some future occasion the commitment-limited restraints that have governed Soviet conduct in respect to the Middle East might break down. It is, for example, possible that the Soviet Union might mismanage a crisis, maneuvering itself into a rigid commitment that would at once enhance its bargaining strength and at the same time entail a real risk of war with the United States. Such Soviet conduct, however, would be a clear departure from the normal pattern of Soviet policy, which has been notably cautious in the Middle East as far as commitment-creation is concerned (and reckless in respect of crisis stimulation, military supply to clients, and propaganda support).

It may therefore be concluded that the Soviet naval forces in the Mediterranean cannot in present circumstances negate the suasion potential of the Sixth Fleet. So long as the Soviet Union cannot persuade political leaders in the area that its naval power will be brought to bear to negate the capabilities of the Sixth Fleet (by overt attack, it is presumed), local leaders will continue to be encouraged (supportive latent suasion) or deterred (coercive latent suasion) by their own perceptions of the power of the Fleet. And, in conjunction with other instruments of statecraft, the Sixth Fleet will also continue to be a valid instrument of active suasion.

We reach finally the third question: if the Soviet Fleet cannot negate the suasion potential of the Sixth Fleet can it perhaps offset its effect through a suasion potential of its own?

A distinction can be made between active and latent suasion. The latter describes the supportive or coercive effects tacitly elicited by others' perception of military power. The former, on the other hand, describes specific supportive or coercive effects evoked by deliberate moves, i.e., political "tactics" such as directed deployment, force-level changes, display maneuvers and so on. In the case of moves intended to evoke such specific effects, it was recognized that the relationship between threats made or encouragement given, on the one hand, and actual tactical openings, on the other, would have to be fairly direct.

There is no doubt that the growth of the Soviet surface navy, and, to a much lesser extent, the modernization of Russian submarine and naval-air forces, have heightened the perceived naval power of the Soviet Union. This, together with an even more dramatic increase in the perceived strategic-nuclear power of the Soviet Union, has undoubtedly increased the latent suasion evoked by the Soviet Navy. But this capacity for latent suasion is not likely to be reinforced by a record of successful active suasion. Given the ultimate relationship between tactical capabilities that are both important and usable on the one hand, and the successful exercise of active suasion on the other, the shortcomings of the Soviet Fleet as a combat force inevitably restrict its political utility.

Today's Soviet surface fleet, like the submarine and naval-air force of the past, has an impressive sea-denial capability. For the time being, though not for long perhaps, the Soviet Fleet could even challenge the Sixth Fleet in direct combat, if only in ideal circumstances, i.e., the "splendid" all-out missile strike. However, no vainglorious Soviet claims, and no amount of exaggeration, can endow this narrowly specialized sea-denial force with the capabilities required to coerce or support the littoral states of the Mediterranean.

The Sixth Fleet can support friendly states by providing air defense for them with its fleet-defense aircraft; it can also provide air attack capabilities, including all-weather attack capabilities absent in local air forces. And the Sixth Fleet also has in its Marine unit a small but versatile organic ground force which can be rapidly reinforced. For coercive purposes as against third parties the Sixth Fleet has all the above capabilities as well as the Marine capability for forcible entry by amphibious means. At sea, the Sixth Fleet can protect shipping or deny passage; it can blockade and it can force a blockade, whether imposed by coastal artillery or by forces afloat.

Soviet naval forces, by contrast, have no coercive or supportive air capabilities (let alone specialized all-weather, ECM support, or reconnaissance capabilities), while Soviet amphibious forces are small and lack a significant opposed-landing capability. This leaves the Soviet Navy with only the sea control options and perhaps a blockade-breaking capability (if Soviet SSMs are sufficiently versatile for use against ground targets, such as coastal artillery emplacements).

Examining likely conflict situations, we find that, in support of Syria or Egypt against Israel, the Soviet Fleet could only attack targets that are of secondary importance in the Arab-Israeli conflict, i.e., commercial shipping; it could not intervene usefully in the decisive combat on land and in the air. It is also apparent that in the event of a successful Israeli offensive, the Soviet Fleet could not protect either Egypt or Syria, for its SAMs would add very little to coastal air defenses, while its SSMs have no significant capability against land targets unless provided with nuclear warheads. The same limitations apply to supportive operations on behalf of Yugoslavia, or to coercive operations against it for that matter. And they also apply to coercive operations against NATO members, Turkey, Greece, and Italy.

With the noted exception of sea control and coastal defense capabilities, the poverty of Soviet sea power is apparent, and it is inherent to its configuration as a highly specialized sea-denial force. Unless future developments make the purely maritime dimension of Mediterranean conficts much more important than it now is, e.g., in a future Greek-Turkish dispute over the Aegean, the political utility of the Soviet Fleet must remain as restricted as its tactical flexibility. Nor is this state of affairs yet changed by the commissioning of a couple of VTOL carriers: a naval-air force of 20 or 40 small fixed-wing aircraft will not have much political impact in the over-militarized Mediterranean. A larger carrier fleet, perhaps with bigger carriers, would be quite another matter.

Regional Constraints

If the theoretical grid of the suasion possibilities of the Sixth Fleet emerges virtually unscathed when confronted with the reality of Soviet sea power, this does not mean that its full potential can therefore be realized. In calculating the degree to which American naval power can effectively support friends and coerce enemies, factors more important than the Soviet Fleet must be taken into account. These factors are, first, the cultural-political orientation of the littoral states towards each of the super powers, and, in particular, their propensity to be influenced by acts of armed suasion; and, second, their military capability.

Unless units of the fleet are to serve as mere symbols, e.g., as strategic "tripwires", it is clear that the political impact of particular threat or support maneuvers will be a function of the *saliency* of the specific capabilities to which they are related in the wider context of all relevant locally deployed forces. A littoral state endowed with a powerful air force cannot be decisively supported (or coerced) by the implied support (or threat) of the air power available on the decks of a pair of CVs. Similarly, the threatened intervention of two battalions of Marines might be decisive in the setting of a Lebanese political crisis but it could not serve to support or coerce one of the several Middle Eastern powers possessed of powerful ground forces.

Before comparing the inventory of local military capabilities with the inventory of the Sixth Fleet's capabilities in order to determine the saliency of the latter, we must address the more recondite factor of cultural-political orientation, at least insofar as the Arab states of the southern shore are concerned. As it has been elsewhere argued in some detail (Luttwak, 1973:3-9), national propensities to respond to armed suasion vary considerably, and such propensities may in the end matter more than any other factor, including the balance of armed power.

The great change that has taken place in the Mediterranean region since the 1950s has been the emergence of the countries of the southern shore as effectively independent states. This did not necessarily coincide with the formal termination of colonial rule. Only in the case of the Algerian "provinces," Tunisia, and Syria did the formal transfer of power mark the attainment of a genuine independence. Elsewhere, formally independent states became independent in reality only much later, when the organic links of dependence between local ruling elites and British power were broken. Only in Algeria was there an actual struggle for independence, against the power of France. There was conflict in Libya, Egypt, and Iraq also, but in those countries the struggle was waged between different local elites: one traditional, or upper-middle-class commercial, and Western-oriented, and the other lower-middle class, military-bureaucratic, and intensely anti-Western. These struggles were won by the nationalists, but not until several years after the formal grant of independence. In the case of Egypt and Iraq a full generation thus intervened between formal and actual independence.

The fundamental reason for the victory of the lower-middle class nationalists over the traditional (and upper-middle class) Western-oriented elites was the ability of the former to mobilize the masses. The nationalists could appeal to the Islamic sentiment and the xenophobia of the masses while their opponents could not.

Given this background, it is highly inaccurate to portray the Soviet presence in the Middle East as having "replaced" former Western suzerainties. The two phenomena are very different: the Western suzerainties had the fixed political meaning of subjection; the Soviet presence does not. Whereas the British and the French had a wide measure of control over the entire political life of the dependent countries, the Soviets have only a limited influence on external policy which is liable to wane abruptly (and then perhaps swiftly increase again).

Ruling groups that acquired their power by satisfying the xenophobic desire of the masses for the exclusion of foreign influence, and by catering to the Islamic demand for the exclusion of unbelief and all unbelievers, cannot proceed to develop close and permanent links with the Soviets, themselves as visibly foreign as the British or the French and certainly unbelievers. Nor can these powerful popular sentiments be outweighed by the undoubted leftism of many of the more educated members of the new elites. The victor in the struggle against the British and the French has not been the Soviet Embassy but the Mosque, even though the influence of religious leaders as such is very small—precisely because the nationalists have managed to capture their role as upholders of the faith in the political arena.

It is for this reason that the presence of Soviet military and economic operatives and the inflow of Soviet equipment, military and industrial (including some Soviet capital), has never resulted in Soviet political penetration to a degree at all comparable in depth and resilience to that achieved by the former suzerain powers. And it is for this reason that the region no longer offers the opportunities for American political influence that were once present. In particular, the exercise of American naval suasion through the Sixth Fleet can no longer be consolidated by installing client-leaders in the arena of local politics. Similarly, local policymaking has become refractory to direct and overt external pressure. As a result, even successful military moves may turn out to be politically sterile if not actually counter-productive. The classic example is of course the 1956 Anglo-French intervention against Nasser's Egypt.

The goal of that operation was not to occupy the Canal in order to open it for international transit, but rather to destroy Nasser's regime. This was to be achieved by placing military forces in Egypt that would defeat the Egyptians, destroy the authority of the military junta, and thereafter act as the arbiters of Egyptian politics. It was imagined that Nasser's regime would be discredited by defeat and that a rival leadership of the traditional type would then emerge, a leadership that would seek Anglo-French support, and make this support the pillar of its own power.

When politics were still monopolized by the former elites, this sort of operation was perfectly feasible; indeed governments were changed during World War II in both Egypt and Iraq precisely through the agency of direct British military action. Unaware of the fundamental change that the mobilization of Islamic feelings and the xenophobia of the masses had wrought, the British government attempted to repeat in 1956 an operation on the lines of the removal of Rashid Ali's pro-German regime in Iraq in 1941. It is significant that Eden acted at the direct instigation of the leading practitioner of the old politics in the Arab world, Nuri es Said of Iraq.

The Suez episode made it clear that in Egypt at least the forcible expression of Western desires could no longer make or unmake local governments. If the British had persevered they could no doubt have occupied Cairo and installed their own man in the presidential palace. In the past this would have sufficed to make the new regime the accepted government of the land, and the support of a foreign military power would have made his position all the more solid. By 1956, however, any visible foreign support made rulers of the Middle East ipso facto illegitimate. As a political operation the Suez adventure was doomed from the start; the fact that it was interrupted halfway through by American intervention in no way affected an outcome that was preordained. Nasser's subsequent record of mounting defeat and increasing popularity revealed how futile an Anglo-French victory would have been in 1956.

The urban society of Egypt is much more sophisticated politically than the regional norm; in 1956 there were still Arab countries where external force could be highly effective in the arena of local politics; there were even places where rulers could be made or unmade by direct and overt outside intervention. Indeed in the case of the Trucial Sheikdoms this was still true in the early 1970s. But there is no doubt that Egyptian politics set the tone for Arab political conduct as a whole, and anticipated the shape of things to come.

The implications of this structural change in the nature of domestic politics within the major Arab countries are clearly negative for the exercise of naval suasion. First, it is no longer possible to make lasting impact on the conduct of a local power by exerting direct control of its governance (as opposed to influencing the behavior of its government)—a foreign power can no longer install its own candidates as the rulers of the land and expect that their rule will survive. Second, the effectiveness of naval suasion in influencing the conduct of governments is also reduced. The logic of domestic politics is such that leaders would prefer to defy foreign demands, however reasonable and discrete, since the gains of overt defiance, to the leaders themselves, are greater than the costs of defiance to the nation. A nationalist leader engaged in a continuing struggle to build a base of popular support can exploit any exercise of naval suasion which is hostile (or which can be so portrayed), to seek the approval of the masses. The costs of defiance are costs to the nation as a whole; its gains are gains for the ruling group exclusively.

Thus, when forced to choose, the regimes in the area prefer to accept military setbacks at the hands of foreigners rather than comply with their demands, however moderate, because the latter would weaken their support from the masses. A very clear example of this logic was evident in Egyptian policy towards foreign aid during the 1960s. On several occasions the Egyptian leadership deliberately provoked interruptions in aid flows from the Soviet Union as well as the United States, calculating each time that the loss of tangible values to the nation counted for less than the political gains to the regime of being seen to strike postures of defiance.

Nor is there anything exotic about such attitudes. The preference for the irrational—in the formal sense of a failure to align material means with material ends—defiance of foreign demands is as widespread as the variations of the phrase, "millions for defense and not a penny for tribute." A penny, after all, is not a high price. But, of course, there is a difference of degree: regimes which came to power by appealing specifically to popular xenophobia are apt to be particularly irrational (from the national viewpoint) in their preference for defiance over compliance when faced with the threat of foreign military action.

Nor, on the other hand, was there anything particularly craven or unpatriotic about the members of the former ruling elites who were ready to give preponderant foreign power its due. Their willingness to comply with the demands expressed through the exercise of armed suasion did not result from personal cowardice or elite corruption but rather from a worldly realism which was far from despicable. Being informed of the greater costs of defiance to themselves and to the nation at large, and being dependent on outsiders for their own power, they saw no purpose in resistance merely in order to strike empty poses of defiance. Hence, their conduct appeared rational to outsiders.

By contrast, for the new ruling groups, whose power-base derives from their appeal to non-Westernized masses, poses of defiance are not empty at all. With mass opinion neither worldly nor informed, and highly emotive, it makes perfect sense to engage in political behavior which appears to be irrational to outsiders. Such behavior would only be truly irrational if the goals pursued were national and material. They are neither. In practice, "economic development" is a low priority goal for all but a few of the regimes of the Third World. The accumulation and preservation of political power comes first.

The adverse change in the political climate of the Middle East, which has made it so refractory to external pressures, has been compounded by a natural concomitant: a rapid growth in local military power.

In the 1950s, the carriers of the Sixth Fleet offered a range of tactical counter-air and attack capabilities that were potentially decisive in the regional context. Given the small and unsophisticated air defenses of the littoral states, the attack aircraft of the carrier air wings could inflict heavy damage upon both military and civilian targets with virtual impunity. Similarly, the fleet defense fighters embarked on the Sixth Fleet's carriers were vastly superior in quality and not numerically inferior to those of local forces. Until 1956 even the best of local air forces, the Israeli and the Egyptian, still relied on piston-engined fighters (Spitfires, Mustangs, and Mosquitos) deployed alongside one or two quadrons of first-generation jet fighters (Luttwak, 1975: 123-126).

Thus, the carrier air wings of the Sixth Fleet, with their superior aircraft, could protect the skies of friendly nations and also operate freely in an offensive counter-air role over all the territory within their range. At the same time, the carriers had very little to fear from local air power and could operate safely even very close to hostile shores. This is no longer so. The growth in local air defenses has been spectacular, and would severely constrain operations over Egypt, Israel, and Syria. Modern Soviet SAM systems have been deployed in parts of the region in a density found nowhere else in the world except for certain air-defense districts in the Soviet Union. American naval attack aircraft no doubt retain a qualitative edge over local air defenses owing to their advanced ECM and precision-guided munitions, but the net striking power of the carriers has been sharply reduced. For this reason carrier-based attack capabilities are no longer as salient as they once were in the minds of local leaders.

Similarly, the counter-air capabilities of the Sixth Fleet, once so superior in quality that all numerical comparisons were redundant, no longer have this absolute advantage. Navy pilots are still undoubtedly much superior to those of most local air forces. Navy F-4Js are no doubt better than Israeli F-4Es, which in turn have some small advantage in air-to-air combat over the MiG-21 Js in Arab hands. When F-14s reach the Sixth Fleet carrier wings, this qualitative margin will be further enhanced. At the same time, Egypt can now deploy two or three squadrons of MiG-23s as well as some 250 MiG-21s and Mirage IIIs; Israel can deploy more than 200 F-4s and Mirage-type aircraft; and even Syria has an operational force which includes more than 150 MiG-21s and some MiG-23s. In the 1950s, by contrast, there were more operational first-line jet fighters on board a single U.S. carrier than were to be found in either the Egyptian or Israeli air forces.

This decline in the relative air capabilities of the Fleet has reduced their saliency, and it has therefore reduced their value for the purposes of action suasion. Furthermore, this applies not only with respect to the capabilities of potential adversaries, but also to friendly states. In the former case the relative decline diminishes the effectiveness of the Fleet's coercive active suasion; in the latter case its supportive suasion is similarly affected.

Partly because of this adverse change in the air power balance, and partly because of the consolidation of local ground forces, amphibian "projection" capabilities have also lost much of their former saliency. For example, it would no longer be possible to intimidate Egypt or even Syria by threatening an amphibious landing, even if an entire divisional-sized Marine amphibious force (MAF) were involved. Nor could friendly powers in the Levant be given an instantly overwhelming advantage through the introduction of Marine forces, even of divisional size. It is true that an MAF could give very powerful assistance even to the most considerable military power in the area, but on its own it would no longer represent the utterly decisive force it once would have been.

In the 1950s, no army in the area was equipped with weapons that were of first-line quality by world standards. Indeed, until the mid-1960s the bulk of the ground-force equipment deployed in the region was of World War II vintage: towed 25 pounders and M.38 122 mm Soviet howitzers in the artillery, T-34 or Sherman tanks in the armor units, and small arms of similar vintage. There was some equipment of more recent origin but it was in short supply. More important, until the early 1960s none of the armies of the area were able to attain standards of efficiency and discipline comparable to those of Marine troops. Now this is no longer so. One local army, the Israeli, has clearly exceeded U.S. standards in its better units, while the Egyptian and Syrian forces, for all their tactical shortcomings, have attained quasi-European standards at least in defensive operations.

All this does not mean that the "disposable" (i.e., net of self-defense) capabilities of the Sixth Fleet can now be negated by local military forces. If that were so, the political utility of the Fleet would now be very small, since, aside from serving as a tangible symbol of threats of promises to be enforced by non-naval means, the Fleet could not serve as an effective instrument of armed suasion. For all the time-lags and errors of perception, a force that is not expected to be able to contribute significantly to the defense of friends and the defeat of enemies would not serve to encourage the former or intimidate the latter.

It has not yet come to that. The "disposable" defensive and attack capabilities of the Sixth Fleet have not in fact been negated by local military forces despite the indubitable reduction in their saliency. For this there are several explanations. First, the Fleet retains its net and uncontested superiority over local armed forces as a sea-control force. Second, it continues to enjoy a significant lead in advanced and specialized air capabilities. Third, the level of armament found in the case of Egypt, Israel, and Syria is not yet the norm for the southern shore as a whole. Finally, in the NATO context the Sixth Fleet remains a key source of tactical-nuclear support, a capability which is of particular importance in the Mediterranean sector of the Alliance.

What remains of the Sixth Fleet's suasion potential may be revealed by reviewing the feasibility and saliency of its major force options against—or on behalf of—the individual countries of the region.

The survey begins with Morocco in the Arab far west. Morocco is still ruled by a king and is still Western-oriented, though now edging towards non-alignment. Its military establishment is small in relation to its population by Middle-Eastern standards, though not of course by African standards. As far as supportive operations are concerned (and therefore supportive suasion), the level of Moroccan air power capabilities is such that even a single naval air wing could provide more and better counter-air than that which the local air force could provide (The Military Balance: 1978-1979:40). Similarly, the attack capabilities of a single carrier wing far exceed those of the local air force, (two squadrons of Magister light strike aircraft). As for ground forces, only a full Marine division could have a decisive impact alongside Moroccan ground forces, which include 50,000 men equipped with more than 200 light and medium tanks. All air-projection options would be feasible and significant with regard to offensive operations. The forcible entry of Marine forces of divisional size would also be feasible, though sustained inland operations with a single MAF would have to be ruled out. In sum, for Morocco the Sixth Fleet retains the tactical ascendancy required for the effective exercise of suasion.

Algeria, with a population of similar size, has a more substantial air force, which includes two squadrons of MiG-21s and roughly 90 strike aircraft: MiG-17s and some Su-7s (The Military Balance 1978-1979:35). The air power of two carrier air wings should still be decisive in counter-air operations, whether supportive or, more likely, offensive. Given the absence of elaborate missile air-defenses, attack operations would be relatively unimpeded while the Fleet's attack capabilities could more than double local capabilities in a supportive mode. Unless a substantial part of the Algerian ground forces were already absorbed by an ongoing conflict (e.g., with Morocco), even a full MAF-sized landing would involve high risks, given the presence of substantial local armor, including 300 T-54s/T-55s; in any case inland operations by a single MAF would not be feasible. On the other hand, as far as supportive operations are concerned, a Marine division would make a substantial contribution to Algerian strength on the ground.

Tunisia is much less heavily armed than any of its neighbors (The Military Balance 1978-1979:43). Given the weakness of the Tunisian air force and the lack of significant air defenses, Sixth Fleet air capabilities would be decisive for supportive or offensive operations, whether for counter-air or for attack missions. Given the size and strength of Tunisian ground forces, wholly lacking in combat experience, even a brigade-sized Marine landing force would outweigh local capabilities, whether in contributing to Tunisian ground defenses or for offensive purposes.

Libya's population is half the size of the Tunisian and one-eighth as large as that of Morocco or Algeria. On the other hand, vast oil revenues have been freely expended for the purchase of arms in recent years (The Military Balance 1978-1979:40). These weapons are not, however, effectively integrated into trained and organized combat forces. Libyan military personnel lack operational training, technical education, and maintenance discipline. Command cadres are short of organizational and leadership experience. The Sixth Fleet's air power should therefore remain decisively superior in all types of operations. As far as ground capabilities are concerned, it is not thought that the divisional-sized Libyan armored force is an effective combat unit as yet, and this ought to ensure the tactical ascendancy of Marine forces of divisional size, as well as the feasibility of forcible entry.

With respect of Egypt, only specialized capabilities, for example ECM and reconnaissance, would be of any significance. Like the other countries of the North African shore, Egypt does not have naval forces capable of detering the close approach of Sixth Fleet units. Otherwise all the major force options are foreclosed, with the partial exception of counter-air operations, for which the aviation of two wings would afford a significant—though not decisive—capability.

Israel has a population not much larger than that of Libya and one-tenth that of Egypt, but its military establishment is hyper-developed. The Sixth Fleet would have no significant, let alone decisive, offensive air options against Israel short of nuclear attack. As far as supportive operations are concerned, Fleet F-14s could add a significant qualitative edge to Israeli air defenses, especially against special threats such as low-level penetration by SU-19s or high-altitude intrusion by MiG-25s. On the other hand, it is unlikely that much "disposable" counter-air capability would be available to help the Israelis in a major conflict situation because a full CAP deployment would probably be needed for the defense of the Sixth Fleet itself. As far as attack capabilities are concerned, A-6 type aircraft and especially the EA-6Bs could offer very valuable specialized support to Israeli air operations. Even a full Marine division would not have a decisive impact on land, though its support would be valuable in an all-out war, especially if its arrival on the scene were timely.

Lebanon still offers a clear field for Sixth Fleet capabilities. Fleet aviation would be decisive in both an offensive or supportive mode, both in the counter-air and in the attack role. Even a brigade-sized Marine force could achieve forcible entry; alternatively, it could provide supportive ground capabilities exceeding those of the Lebanese Army.

Syria, with twice the population of Israel, has a vast arsenal of Russian equipment, featuring, among much else, 200 MiG-21s and roughly 2,000 modern battle tanks as well as extensive air defenses (The Military Balance, 1978-1979:42-43). Nevertheless, Sixth Fleet aviation could be decisive in counter-air operations for supportive or for offensive purposes. On the assumption that the bulk of Syrian air defense forces would be concentrated on the Israeli sector, attack options would also be available offensively. On the other hand, unless the Syrian Army was already engaged in combat on the Israeli sector, even an MAF-sized force could not be usefully introduced into Syria.

Turning to the countries of the northern shore, we find that their generally higher level of armament means that the Sixth Fleet, even if much reinforced, could rarely play a decisive role.

Spain, which is outside NATO but has a bilateral security treaty with the United States, is remote from overland dangers but not from maritime threats. Spain has a significant naval role in Western strategy, to some extent in the Atlantic, but much more significantly in the Mediterranean. With two carrier air wings, the Sixth Fleet could add substantially to Spanish counter-air capabilities and would have almost twice the attack capability of the Spanish Air Force. However, land-based USAF units would also be available in any contingency in which the Sixth Fleet itself were engaged. As for ground forces, nothing less than a full Marine division would add significantly to Spanish capabilities. Since the contingency of a Spanish-Portuguese conflict in which Portugal served as the base for Soviet military power is exceedingly remote, the political significance of these Navy capabilities is also of secondary importance.

Sixth Fleet capabilities would similarly be of importance to Italy only in the event of a major external threat emanating, say, from the Yugoslav frontier, especially in a situation in which France remained neutral. In that eventuality, the Sixth Fleet's sea control would be very important in ensuring strategic access to Italy.

Yugoslavia faces the certainty of a succession crisis, and one that may well be compounded by the ethnic conflicts latent below the surface (Luttwak, 1974b:43-51). Either eventuality (and both for a certainty) would offer opportunities for Soviet intrusions, certainly political but possibly military also. If Italy made her territory available for supportive operations on behalf of Yugoslavia, the importance of Fleet capabilities would be marginal. If Italy had by then undergone such drastic political change that her position was actively hostile to supportive American operations, then no naval deployment would be tactically feasible in the narrow Adriatic sea. If Italy were neutral, on the other hand, and NATO membership would not absolutely require Italian cooperation with the United States in the defense of Yugoslavia's integrity, Sixth Fleet capabilities could be critical in ensuring strategic access to Yugoslavia.

Further, there would also be a role for the organic air capabilities of the Fleet. The only counter-air forces now available to Yugoslavia are some eight squadrons of MiG-21 fighters, and support from Fleet carriers could be significant especially if air-defense coverage over Yugoslav territory obviated the need for CAP coverage for the carriers themselves (The Military Balance 1978-1979:32). The Fleet's attack capabilities would be more significant still, since all Yugoslav ground-strike aircraft are clear-weather/light-weight trainer conversions. In view of the nature of the terrain, their declared guerilla strategy, and the substantial size of Yugoslav ground forces, Marine capabilities would be useful only for specialized or symbolic purposes.

Albania is neither in NATO nor the Warsaw Pact, and it is also the smallest, poorest, and militarily weakest country in Europe (the mini-states excepted). It should offer a clear field for all air options both for offensive or supportive purposes—in the unlikely event that the latter would be appropriate. Terrain and a (non-political) predisposition to guerilla warfare on the part of the local population rule out offensive ground operations beyond the coastal zone. In a supportive role, on the other hand, specialized U.S. ground-support role forces could magnify Albanian self-defense capabilities. There should be no real obstacle to forcible entry by Marine forces for a limited coastal occupation.

Greece is a front-line member of NATO with a particularly difficult border sector facing Bulgaria. For geographical reasons alone, Sixth Fleet counter-air support could be decisive in Thrace, if, that is, CAP requirements do leave a disposable surplus available for the defense of Greek airspace. Fleet attack capabilities would also add substantially to those of the Greek Air Force, which are limited by aircraft quality as well as numbers. However, land-based USAF and NATO air power would also be available in any contingency in which the Fleet was engaged. The introduction of Marine forces could be of considerable significance; a complete Marine division air wing team would be at least equivalent to one-fourth of the Greek Army, excluding armor, in terms of ground capabilities, and more significant still as compared to Greek air power.

Turkey has a large army but a relatively small air force with limited counter-air capabilities and no all-weather attack capabilities at all. Fleet counter-air supportive operations could be very helpful if the aircraft could be spared the task of carrier self-defense. Attack capabilities should be helpful in any case even if the contingency were to entail general naval warfare with the Soviet Union.

The results of this brief and cursory review are summarized in Tables 2 and 3, where the significance of selected Fleet capabilities is tabulated country by country. The potential significance of Fleet operations

TABLE 2

Tactical Viability and Significance of Fleet Operations in Respect to Individual Countries on the Southern Shore of the Mediterranean

	MOROCCO	ALGERIA	TUNISIA/LEBANON LIBYA	SYRIA	EGYPT	ISRAEL
Offensive Operations						
Counter-air	Decisive	Decisive	Decisive	Significant	Significant	None
Attack	Decisive	Decisive	Decisive	Significant	Symbolic	None
Forcible entry	Significant	Symbolic	Decisive	—	—	—
Ground force	Significant	Symbolic	Decisive	Symbolic	—	—
Supportive Operations						
Counter-air	Decisive	Decisive	Decisive	Decisive	Decisive	Significant
Attack	Decisive	Decisive	Decisive	Decisive	Significant	Significant
Ground forces	Significant	Significant	Decisive	Symbolic	Symbolic	Significant
Specialized Air Support Ops*	—	Significant	—	Significant	Significant	Significant

*In the presence of substantial local forces, this category applies to specialized capabilities such as ECM support, ELINT etc. Such capabilities *ipso facto* cannot be decisive.

TABLE 3
Tactical Viability and Significance of Fleet Operations in Respect to Individual Countries on the Northern Shore of the Mediterranean

	SPAIN	ITALY	YUGOSLAVIA	ALBANIA (Offensive & Supportive)	GREECE	TURKEY
Supportive Operations (1)						
Counter-air	Significant	Symbolic (2)	Significant	Decisive	Decisive (5)	Significant
Attack	Significant	Significant	Decisive	Decisive	Decisive	Significant
All-weather attack	Decisive	Decisive	Decisive	Decisive	Decisive	Decisive
ECM support	Decisive	Significant	Decisive	Decisive	Decisive	Decisive
Ground force	Symbolic	Symbolic	Symbolic	Significant (3)	Symbolic	Symbolic
Forcible entry	—	—	—	Decisive (4)	—	—

1. Offensive also, in the case of Albania.
2. It is assumed that land-based USAF and NATO air power would also be available; this, together with local capabilities, would make Fleet counter-air support of marginal significance.
3. Special capabilities, e.g., anti-tank weapons, would be important in a supportive role.
4. For occupation of the coastal strip only; terrain and population factors inhibit inland operations.
5. Only for the N.E. sector and only then if Fleet self-defense leaves a disposable counter-air surplus.

is roughly classified under three categories: decisive, where Fleet capabilities exceed by far those of local forces; significant, where a substantial and visible contribution to local defense could be made by supportive Fleet operations or a meaningful threat could be presented; and finally, symbolic, which applies to both offensive and supportive operations, is self-explanatory. If the tactical effects are assessed as insubstantial, or the reference is not applicable, the item is left blank.

The relationship between the saliency of the Sixth Fleet's capabilities and the Fleet's potential as a political instrument is not, however, direct or even proportionate. Perceptions of combat capabilities in the minds of the political elites of the Mediterranean region are not likely to be highly accurate, but there is nevertheless bound to be some ultimate connection between the *net* difference that the Fleet's intervention would make to the local balance of power on the one hand, and the degree to which local leaders can be encouraged or coerced through the exercise of active suasion on the other. The data of Tables 2 and 3 can therefore be read as a statement of the necessary (but not sufficient) conditions for the political uses of the Sixth Fleet (Luttwak, 1974a:74-79).

Conclusions: The Sixth Fleet and National Strategy

Under whatever name, for some years yet the Vietnam Doctrine will guide the general conduct of American foreign policy. With brutal simplicity the doctrine can be summarized as an attempt to retain all these American interests worldwide that can be protected without significant new military deployments. It is therefore, not a strategy of retrenchment; the *latter would entail the willingness to abandon significant interests.* Instead, American interests worldwide are to be defended by means of non-military instrumentalities of statecraft, and *new* ground-force deployments are specifically excluded.

It is understood that unless overseas interests are abandoned, the defense of those which are territorial in nature cannot be a matter of static "commitments." Indeed, it is misleading to think in terms of commitments at all. The word suggests a one-time acceptance of some fixed obligations, instead of the reality of continuing interests that require a continuing protective effort as variable as the hostile pressures that threaten them. The ground rules of the doctrine merely reflect that post-Vietnam opposition of domestic opinion to any new American troop deployments.

In the event of an overt and major Soviet attack upon Western Europe it is generally assumed on all sides that the machinery of NATO planning would begin to function automatically, and that forces in the United States earmarked for SACEUR would begin to redeploy across the Atlantic equally automatically.

Since the legitimacy, rationality, and utility of this prospective redeployment is not yet being questioned by authoritative voices in the United States, the deterrent effect of the American contribution to NATO ground reinforcements after M-Day—with all that this entails in the linkage to strategic-nuclear deterrence—remains unimpaired. Elsewhere the situation is less clear.

It is not, for example, certain that in the event of a North Korean invasion of South Korea there would be sufficient congressional support for major reinforcement to the one division still deployed there. The presence of this division already in place is, of course, a major deterrent on its own, if only because of the known difficulty of extricating it in the event of a general attack from the north. But this particular deterrent operates only at the maximal level of threat. If the North Koreans were to launch an attack on the sector held by ROK forces only, and if they were successful in advancing, the United States would have to choose between withdrawing its one division, exposing it to encirclement, or reinforcing the ROK sector with American troops. At that point, it seems quite likely that there would be a strong congressional pressures—pressures with legal teeth—against the deployment of additional ground forces to Korea.

In the light of this, the deterrence of North Korean adventurism is contingent on the capability of the local South Korean forces—and not American ground troops—to contain an all-out attack on a selective front, albeit with U.S. air and naval support.

In a less stark manner, similar considerations are operative in the case of Mediterranean interests vulnerable to outright military attack, or to political pressures. The one exception is the contingency of an intervention in the Persian Gulf. In the event of an oil embargo which left Western Europe and Japan without adequate fuel supplies to sustain essential production, the United States is likely to mount an operation to occupy the oil fields of the Arabian Peninsula. Given the absolute dependence of Western Europe and Japan on oil produced in the territory of Arab states (i.e., given the insufficiency of non-Arab oil), the threat of intervention is the *only* deterrent to actions on the part of the oil producers whose consequences could be catastrophic. With this one exception, the major interests which may require the commitment of U.S. military power have a generally territorial character, and their protection would have a clearly defensive character.

It is apparent that as long as the post-Vietnam rules remain in effect on the external policy of the United States, an added burden is placed on the "capital-intensive" instruments of military power, the naval, remotely based air, and strategic-nuclear forces. This applies not only to the actual combat deployment of military forces, but also to the political application of military force. To the extent that the armed suasion of American ground forces is ineffectual, because their commitment in actual combat is not credible, such suasion must be generated by remotely-based air and naval forces operating on their own, or more likely, in support of non-American ground forces. To the extent that the threatened, or offered, intervention of American ground forces is not effective to support or coerce because of the known opposition within the United States to intervention, support or coercion must rest on air or naval forces, i.e., forces that are not affixed to the locale of combat.

But the fact that American ground forces are now less available for redeployment than air or naval forces does not mean that their capabilities can in fact be freely replaced with naval or remotely based air forces. For one thing, there may not always be allies at hand ready to contribute ground forces to fight for American interests—even if these are also alliance interests. This applies, for example, to the very important non-territorial interest of ensuring the continuity of oil supplies. Even where interests are territorial, and where there are locals technically able to fight on the ground, the division of labor envisaged by most operational interpretations of the Vietnam Doctrine raises very serious problems for alliance diplomacy. A strategy that is acceptable in the arena of American domestic politics because it places neither American territory nor many American lives at risk may be unacceptable in the arena of interalliance politics for exactly the same reasons. Such a strategy would certainly be unacceptable to the European members of NATO.

Aside from the objections to a division of labor within the Alliance, in which the Europeans contributed the cannon-fodder and the Americans contributed high-technology air and naval forces, there is a strategic objection also. Since European territory and European populations would be exposed to mass destruction in the course of an active defense against a Soviet offensive, the Europeans have always placed much greater emphasis on deterrence than on defense. And the best guarantee that the United States would actually resort to nuclear retaliation for an attack upon NATO Europe is precisely the presence of a large American troop contingent. Therefore, it would not be possible to implement the logic of the Vietnam Doctrine in the case of Europe without prejudicing the very survival of the Alliance.

In the last analysis, the Europeans do have an alternative to the Alliance, and this is to conciliate the Soviet Union singly rather than to deter it jointly with American support. If this alternative is ever adopted, the result would be a neutralization of Western Europe under Soviet control. This would no doubt be unwelcome to many Europeans, but it would also immediately reduce the United States to the role of an isolated North American power, or North American-Pacific power at best. It is in this broad context that the role of the Sixth Fleet as an instrument of policy in the Mediterranean must be defined.

It is apparent that the rigidities imposed by domestic political constraints inhibit local-based forces (both army and air force) to a much greater extent than they inhibit remotely-based air power, and especially naval power. Unless these rigidities disappear, the defense of territorial and quasi-territorial American values by means of armed suasion will have to rely either on ground and local-based air forces already in place, or else on naval forces alone. Since threats are variable and dynamic, it is unlikely that the pattern of ground-force and local-based air deployments inherited from the past will for long continue to coincide with the location of those American territorial interests that come under hostile pressure. It follows that armed suasion will increasingly be naval suasion since only naval capabilities remain freely deployable.

Other factors have compounded the change. The shift in the strategic-nuclear balance from a net American superiority to the present rough parity has naturally increased the burden of non-strategic armed suasion in general. In effect, the post-Vietnam restraints on the deployment of other military forces have shifted a greater share of an already magnified burden upon the navy.

It is unfortunate that these unrelated strategic developments, which affect the "demand" side of the force equation, happen to coincide. It is more unfortunate still that they do so at a time when the "supply" side of the equation is affected negatively by a hopefully temporary decline in the numerical strength of the navy. Moreover, at the regional level there has also been an increase in the demand for U.S. Navy capabilities, owing to the heightened importance of the Mediterranean.

There is therefore a whole accumulation of factors which simultaneously increase the demands placed upon the Sixth Fleet as a political instrument and at the same time make it more difficult than ever to satisfy these demands. Ships, weapons, and supporting systems now under development or under construction should alleviate some of the negative pressures on the supply side of the equation. Until then, it will be up to the national command authorities to resolve any contradictions that may arise between the priorities of tactical prudence, on the one hand, and the requirement of national strategy for a ready instrument of armed suasion, on the other. In the event of a deep Soviet-American crisis, the choice between tactical prudence and the imperatives of national policy will be inexorable. Redeploying the Sixth Fleet from the narrow seas of the Mediterranean into the broad expanse of the Atlantic could be at once an act of tactical wisdom and of political folly. Even quasi-official suggestions that the Sixth Fleet might be withdrawn from the eastern Mediterranean in the event of an intense crisis are in themselves political acts, and costly acts at that. No friend can be encouraged and no enemy can be deterred by a force that will be there at all times except when needed most. The Sixth Fleet may have to live dangerously if it is to serve as a valid instrument of United States foreign policy.

Fortunately, in most foreseeable circumstances it is exceedingly unlikely that any crisis in the Mediterranean region will endanger core values for both sides. Super power relations within the area are very asymmetrical: access to the oil of the Gulf is, for example, a core value for the United States (on behalf of Europe and Japan), while its denial would be a merely desirable goal for the Soviet Union. Equally, the maintenance of a client-regime in Bulgaria is, for example, a core value for the Soviet Union, while its displacement would not have anywhere near the same value for American foreign policy. And when core values are not in dispute, the prospect of a Soviet-American conflict at sea immediately recedes, for its outcome is very likely to be an appeal by the loser to the higher court of strategic-nuclear power, with consequences equally unacceptable to both sides.

The naval confrontation that attended the 1973 Arab-Israeli war was the largest in scale, and by far the most important we have witnessed since the end of World War II. For this reason it is the subject of a detailed case study in the present text. Even within the constraints that derive from the nature of the region's politics, and the further constraints that arise from the growth of local military capabilities, the Sixth Fleet served in 1973 to support effectively a variety of purposes: from the technical, e.g., in providing navigation assistance for the airlift; to the strategic, e.g., reassurance to Israel; to the regional-political, in maintaining the American power-position with friendly Arab states; to the global-political, by the deterrence of the Soviet Union, then seemingly on the verge of contemplating some form of intervention.

The lessons of the episodes are varied, but at a time when theoreticians were increasingly inclined to question the utility of naval power, the case study that follows serves to remind us that the Sixth Fleet remains the most dynamic and versatile instrument of American external policy in a region more troubled and more important than most.

References

Blechman, B., and S. Kaplan (1978) "Force without War." Washington, D.C.: Brookings Institution.

Cline, R. (1974) "Policy without intelligence," *Foreign Policy* (Winter).

Luttwak, E. (1972) *The Strategic Balance*. New York: Library Press for the Georgetown Center of Strategic Studies: 69-75.

———(1973) "The Emergent International System and U.S. Foreign Policy." Stanford Research Institute (SSC-TN-2240-10, June): 3-9.

———(1974a) "American Naval Power in the Mediterranean: Part I: The Political Application of Naval Force." Naval War College, Advanced Research Program, May, 1973. Published in book form as *The Political Uses of Sea Power*. Baltimore: The Johns Hopkins University Press.

———(1974b) "American Naval Power in the Mediterranean. Part II: Mediterranean Undercurrents." Naval War College, Advanced Research Program, April. An edited version has been published as "Issues in the Mediterranean" by the Chicago Council on Foreign Relations, 1975.

Luttwak, E., and D. Horowitz (1975) *The Israeli Army*. New York: Harper & Row.

The Military Balance 1978-1979 (1978) London: The International Institute of Strategic Studies.

Moorer, T. (1975) in Hearings before the Senate Armed Services Committee (5 February, 1974): Part I, Authorizations: 271.

National War College Proceedings (1975) "Defense planning for the future," in the conclusions of the Second National Security Affairs Conference: Washington, D.C.

Zumwalt, E. (1975) in U.S. Naval Institute Proceedings, 100:11:861.

8

American Naval Power in the Pacific: The Policy Environment (1978)

"Eurocentrism" and the U.S. Policy Environment

It is generally agreed that the defense policy of the Carter Administration is governed by a "Eurocentrist" strategy whose impact is of course particularly noticeable in shaping decisions on the composition and deployment of the general-purpose forces. Perhaps less widely recognized is the nonstrategic origin of the Eurocentrist strategy, which appears to be the result of a political or even a philosophical compromise between activists and isolationists, rather than the outcome of a *strategic* assessment of the global costs and benefits of alternative geographic distributions of the American military effort.

Unable to defend full-scale globalism in the wake of the American debacle in Indochina, the advocates of continued activism within the foreign-policy elite staged a deliberate tactical retreat to a perimeter concept, centered on Europe. Unable to carry the day in their moment of greatest opportunity, the years 1973-1975 when military defeat and a grave domestic political crisis were manifest in close conjunction, the advocates of isolationism retreated from the maximalism embodied in the Mansfield Amendment, and reluctantly conceded the continued necessity of a limited American military role overseas, chiefly in Western Europe, with its natural extensions in the Mediterranean.

The resulting political compromise had already emerged in Congress and in the unofficial venues of policy debate by the time that the new administration took office. The Eurocentrist compromise formed the basis of a (new) "foreign-policy consensus," whose absence had been universally deplored during the early 1970s, and which all agreed was an essential precondition of the orderly conduct of policy. Without a consensus, every major initiative of the Executive, from the annual defense budgets to single weapon programs large enough to be visible,

135

were apt to be challenged by arguments proceeding from first principles, which questioned not merely the specifics of implementation but rather the fundamental purposes of external policy as a whole. When the new, Eurocentrist, consensus emerged these paralyzing nondialogues were brought to an end, but the political solution created new strategic problems.

Naturally enough, the compromise between the advocates of isolationism and activism did not reflect any coherent strategic logic. To be sure, the arguments advanced on either side were sometimes rationalized in strategic terms, but mostly they were based on frankly nonstrategic considerations, chiefly being predicated on notions of cultural affinity, or on the presumed common interests of all industrial democracies.

Only by sheer chance could the compromise perimeter defined by the new consensus have coincided with a purposeful strategic arrangement. As it was, the politically defined perimeter of primary concern, i.e. Europe and its extensions, did not in fact correspond to a rational strategic perimeter: the political and philosophical dividing line attempted to separate the strategically indivisible.

Some of the anomalies were immediately obvious. The states of Western Europe did not, and could not, constitute a self-contained strategic zone, being critically dependent on the fuel minerals of Central and Southern Africa. A purely military security vis-à-vis the armed forces of the Soviet Union and its East European clients is therefore a necessary but not sufficient condition of European political security. If the Soviet Union could acquire control of the sources of essential raw materials for which no substitutes exist, it would be able to exercise political leverage over Western Europe even if NATO's defense, and nuclear deterrence, continued to negate the Soviet military threat as such.

This anomaly, created by the geographic discrepancy between the military and the economic dimensions of European security, was soon resolved, not of course by any reduction in the importance of extra-European raw materials for the European economies but rather because of the rapid increase in America's own dependence on these same raw materials. When it became clear that the United States too would be increasingly dependent on Persian Gulf oil supplies, at least until the end of the 1980s, that area was quite easily incorporated in the preexisting American security interest in the Levant, which is anchored in turn on the long-standing U.S. commitment to Israel's security already then a noncontroversial extension of the Eurocentrist compromise.

As for the security of access to the raw materials of Africa it was not until 1975 that the issue reached prominence on the strategic agenda. Until then, Sub-Saharan Africa had been quiescent; African struggles and disorders were already rampant but they did not involve any super power directly. While the fighting between Portuguese, Rhodesians, South Africans, and Black Africans sometimes threatened critical lines of inland communication, there was no concerted *strategic* threat to the supply of nonfuel raw materials as a whole, since the Soviet Union had no active role anywhere in the area. This of course changed quite drastically with the Angolan expedition and the further extensions of Soviet activism that followed in its wake. The Carter administration seemed to be peculiarly inhibited in reacting to the opening of this new front in the Soviet-American rivalry; it was not until the spring of 1978 that the existence of a Soviet threat in Africa was formally recognized in Washington.

But of course the far greater strategic anomaly was the ambiguous role of East Asia in the Eurocentrist scheme. Once Eurocentrism became the common ground of a new consensus between qualified isolationism and retrenched globalism, the obvious dependence of Western Europe on critical raw materials from outside the perimeter could scarcely be overlooked, and the extension of the perimeter was all the more easily accomplished since it was plainly required by specifically American interests also. But the connection between East Asia and European security is of an altogether more subtle order, and its logic though perhaps equally compelling upon close scrutiny is by no means self-evident.

In any case the East Asian anomaly of the Eurocentrist strategy did not emerge in full view until February 1977, when the issue was forced by the official announcement that U.S. ground forces were to be withdrawn from South Korea. Of course it should have been obvious all along that Eurocentrism would by definition entail a strategic withdrawal from East Asia, or at least a considerable reduction in the overall American military effort in the region. After all, East Asia was the only "excluded" zone in which U.S. forces of significant size were to be found. But for three principal reasons the obvious was obscured.

First, in the immediate aftermath of the collapse of South Vietnam it was widely agreed that the remaining U.S. force-deployments and treaty arrangements in East Asia should be frozen at least temporarily, in order to avoid a regional chain reaction which would drive former allies and clients into neutralism or worse. Even those who were convinced of the ultimate wisdom of a total American withdrawal from East Asia, did not wish to see the defeat of American arms in

Indochina immediately translated into wide-reaching diplomatic debacle. This interim consensus on the short-term desirability of preserving the status quo naturally postponed the emergence of the issue.

Second, the fatal incoherence of the American conduct of the war, the resulting disarray in domestic opinion, the deep division and intermittent disorders caused by the war in coincidence with generational and racial turmoils, and above all the profound revulsion generated by the mixed imagery of technological violence and humiliating defeat, all combined to engender a powerful emotional rejection of all Asian things. Within the ranks of the professional foreign-policy elite this was manifest in a willfull inadvertence towards Asia and its security problems. There was a perceptible refusal to confront these problems including the nexus between Asian security and the European balance. Many among the foreign-policy elite were content to simply turn away from Asian concerns, to focus their attention on the far more orderly European scene, where seemingly there were no agonizing choices to be faced and no haunting memories of failure and defeat.

The third reason was not emotional and transitory but in its own way equally obfuscating: Japan itself was treated as an exception in mainstream formulations of the Eurocentrist concept,[1] and this appeared to remedy the most glaring anomaly of all.

The View from Tokyo

Japan's Regional Security Requirements

The largest of all foreign industrial democracies could scarcely be left out of a perimeter which was meant to delimit a commonwealth of the industrial democracies. But the inclusion of Japan however emphatic was, in strategic terms, much more nominal than real. This is so because Japan on its own cannot be treated as a self-contained strategic entity, unless the stability of the adjacent areas, and the security of essential lines of maritime communications, is implicitly assumed as a given. In other words, the addition of Japan to the magic circle defined by the new consensus was predicated on an effortless maintenance of the regional status quo. But strategic arrangements cannot be based on "fair-weather" assumptions; they are supposed to provide *actively* for the preservation of a given status quo, and then to provide for remedial measures in the event of breakdowns in the status quo. A security system which simply assumes the preservation of a secure environment is a contradiction in terms.

Japan's regional security requirements are well known. They have been frequently reiterated by successive Japanese governments over the last two decades and more, and in any case the geography of Japan's physical and economic security makes them fairly obvious.

Korea remains of the essence. Beyond a mere geographic proximity there is in Japan an emotional immediacy towards Korean affairs entirely lacking in Japanese dealings with all other nations, China and the United States included. It is well known for example that the only contingency (other than a direct military attack) which might induce the Japanese to develop nuclear capabilities would be the acquisition of nuclear weapons by the Koreans, North, South, or both.

From Japan's point of view the status quo is highly satisfactory. The Republic of Korea's (ROK) territory is not available for hostile action against Japan while being available for its own protection. In specifically naval terms, it is particularly important that the far side of the Straits of Tsushima is readily available to support combined U.S.-Japanese blockade operations against the Soviet Pacific Fleet. As a bonus, Japan also enjoys a large, growing and profitable trade with the ROK. Moreover, Japan has some relations with North Korea too, though these have been sterile in diplomatic terms while neither important nor profitable in economic terms.

Japanese satisfaction is of course muted by the possibility that North Korea might reopen hostilities, thus bringing all the uncertainties of war to the whole region. Accordingly, it is the official position of the Japanese government that it would view with favor a peaceful resolution of the struggle of regimes, leading to a mutually agreed reunification of the Korean peninsula. On the other hand, Japan's approval of a Korean reunification presumes that a united Korea would be either U.S.-aligned (an exceedingly unlikely prospect) or else very strictly neutral, a condition very difficult to maintain for an isolated small-power adjacent to two giants. It is for this reason that from the Japanese point of view a divided Korea remains preferable to a united Korea, even when the latent risk of war is taken into account. Unofficially, authoritative Japanese point out that a united Korea could also present a direct threat: given the powerful anti-Japanese sentiments of the Korean populations, a unified Korea might always be tempted to strengthen its domestic solidarity by adopting a hostile attitude towards Japan.

At any rate, until such time as a rather problematic unification becomes a realistic possibility, Japan's security requires a secure ROK. And that in turn must mean an American-protected ROK. While it is fully appreciated in Japan that the South Korean economy and demography could both support a military effort fully adequate to deter, and if needs be defeat, any purely North Korean offensive threat, competent Japanese observers stress that a strictly unilateral North Korean initiative is only one of the possible contingencies, and that the ROK also needs protection-by-deterrence against a Soviet-sponsored (or—less likely—a Chinese-sponsored) North Korean attack. Such deterrence obviously cannot be provided by the ROK, no matter how prosperous its economy and how well prepared its forces are.

Japan's security interest in Taiwan is seldom expressed in the explicit (and emphatic) manner in which Korean affairs are discussed. For one thing Taiwan would of course be much less significant as a base for operations against Japan than Korea; any hostile action emanating from that quarter is much more likely to be directed against Japan's southward communications rather than against Japanese territory itself.

Like most competent observers, the Japanese do not believe that an invasion of Taiwan from the mainland is at all probable in the near future. Of altogether greater concern to the Japanese is the possibility that American policy towards Taiwan and Peking might change, and that Taiwan would in effect be abandoned. As far as the Japanese are concerned, the possibility that Taiwan's independence might then be extinguished—serious though this would be if only for economic reasons—would be of lesser concern than the wider implications of such a change in American policy. As the Japanese see it, the American attitude towards Taiwan is a crucial test of long-term intentions. To the extent that the United States seeks to remain over the long term a major power in the Western Pacific, it must retain a network of alliances in the region, of which Taiwan would be part, even if in general U.S. policy runs in a parallel course with that of Peking. After all, over the long term all alignments are apt to change, and parallel action with Peking might give way to equidistance vis-à-vis Moscow and Peking, or even a reversal of relations. On the other hand, if the United States intends to withdraw from the region, however gradually, it would make sense for it to conciliate Peking early, since such a withdrawal would inevitably result in the substitution of a bilateral Sino-Russian balance for the present triangle, and the United States would then be forced to rely almost entirely on the Chinese to contain Soviet activism in the entire region.

Hence the paradox: the Japanese who derecognized Taiwan some time ago now regard the possibility of American derecognition as a very alarming prospect, with far-reaching implications. On the other hand, a formal American "normalization" with Peking which would provide for a contained commitment to Taiwan's security would probably be interpreted as a purely diplomatic adjustment, which need not foreshadow a general American withdrawal from East Asia.

In this context there is a particular contingency which has attracted attention, and some concern, in recent years, though it is not generally regarded as highly probable: the establishment of a Soviet-Taiwanese security relationship (overt or otherwise), featuring the exchange of Soviet security guarantees (possibly tacit) for the use of base facilities on the island. In the wake of treaty-abrogation by the United States, Taiwan's need for a protector and the extremely high value of basing facilities on the island to the Soviets would provide on each side very powerful strategic arguments for arrangement. As against the strategic logic there stands the entire official ideology of Taiwan's regime, in which anticommunism is deeply rooted, and which is ostensibly as hostile to Soviet communism as it is to Peking's version of the faith.

On the other hand, Japanese observers concerned by this particular contingency point out that the Nationalists only became anti-Soviet in order to enlist the support of anti-Soviet forces in their own struggle against the Chinese Communists. Now their anti-Soviet alignment has become almost irrelevant to their position and Taiwan receives very little support on this score. Since the anti-Soviet stance has now become disfunctional it may simply be dropped after a careful preparation of domestic opinion. Of course it is by no means clear whether the Soviet Union would be willing to prejudice its own ideological purity by dealing with the "reactionary" Nationalists, but in this respect a long list of counterideological Soviet initiatives comes to mind, and of course the strategic value of a Taiwan base complex would be very great indeed to the Soviet Union.

It should be recalled in this respect that the Indian Ocean-Western Pacific sea route is of crucial importance for internal Soviet communications, given the volume limits and high costs of overland trans-Siberian transport and the precarious nature of the Northern sea route from European Russia to the Soviet Far East. A foothold in Taiwan would be exceedingly useful to help secure the Soviet military LOC from the Black Sea to the Far East, quite aside from providing base support for offensive Soviet naval and naval-air action, which is at present circumscribed by the severe geographic limitations of the Soviet Far East base complex.

Taiwan and South Korea constitute the inner zone of Japan's security sphere: the connection between the security of the two countries and that of Japan itself is direct. Beyond Taiwan in a southward direction the Philippines, Indonesia, Singapore, and Malaysia are each of considerable economic importance to Japan as export markets, and to a lesser extent, as sources of raw materials. But from a strategic point of view, Japan's intrinsic interest in the internal stability and external security of the four countries is dominated by a more generalized concern over the security of the Japan-Persian oil route. Japanese observers recognize that the more probable threat to oil supplies is at source, in the Persian Gulf producer countries themselves. But they also recognize that in this respect Japan must depend exclusively on the interaction between the producer countries and the United States as well as Western Europe; while Japan's dependence is distinctly more pronounced than either of the latter, it must nevertheless take shelter in the dependence of Americans and Europeans who are far more capable of influencing the course of events in the Persian Gulf. Hence the focus on the security of transit which is a function of regional stability and security in the Western Pacific area as a whole.

Japan's Security and the Eurocentrist Formula

Mainstream formulations of the Eurocentrist formula include Japan in the magic circle of countries which the United States is to help protect by deterrence, and in whose wartime defense it is to participate, should deterrence fail. It is clear that if South Korea, Taiwan, the Philippines, Indonesia, Singapore, and Malaysia are added to Japan itself, then the Eurocentrist formula will have been stripped of its purpose and meaning, which is precisely to disengage from Asia (except Japan), and to focus a diminished American military effort on Europe.

On the other hand, it is equally clear from the foregoing section on Japan's regional security requirements that protection provided for Japan alone would not meet that country's minimum security requirements. The security of Japan cannot be isolated from that of its regional environment since the connection is organic. As it stands therefore, the Eurocentrist formula violates strategic logic; it may be politically palatable in Washington's policymaking environment but it simply does not correspond to a meaningful strategic arrangement: Japan must either be left unprotected or else its essential regional sphere must be protected also. It is perhaps the fact that Japan consists of islands which inspired the peculiar notion of Japan's autonomuous defensibility. But to treat Japan as the exception in the context of a general withdrawal from the region is not far different from undertaking to defend, say, West Germany but not Denmark, the Netherlands, or Belgium.

It is not therefore surprising that it was the sharp reaction of the Japanese government to the initial announcement of the administration's intention to withdraw U.S. troops from South Korea that brought the entire issue of Eurocentrism to prominence. It will be recalled that these initial announcements suggested that all U.S. troops would be withdrawn fairly quickly; that these announcements did not define the postwithdrawal role of U.S. Air Force (USAF) and U.S. Navy forces in Korean defense; and that they made no detailed reference to any compensatory measures, such as accelerated deliveries of military equipment to the ROK's armed forces. Neither did the announcements even attempt to explain the U.S. decision in military, strategic or diplomatic terms: the decision was badly characterized as the fulfillment of an electoral promise.

In addressing the military dimension of withdrawal decision, the administration understandably stressed the economic strength of the ROK rather than the actual balance of forces. It argued that even if the ROK was not now doing so, it could certainly afford to deploy forces capable of contending with the North Koreans. The military *potential* of the ROK was thus compared favorably with the actual military capabilities of North Korea, without reference to the difficulties and delays of translating economic capability into deployed military power, and without reference to the peculiar geographic vulnerabilities of the ROK. Moreover, the military evaluations prepared ex post facto to justify the withdrawal decision were cast in strictly war-fighting terms, and in a strictly bilateral North-South context. There was no sustained analysis of the requirements of deterrence vis-à-vis a Soviet-supported (or less likely, a Chinese-supported) North Korean aggression.

When Japanese officials and domestic critics of the initial decision pointed out that the ROK could not deter an externally supported North Korean attack even with much-expanded military forces of its own, the reply was that the ROK would provide for its own defense, while the United States would continue to remain responsible for the deterrence of externally supported aggression. Aside from general American capabilities, notably including strategic-nuclear capabilities, such deterrence would also be provided by American air power in Korea itself—which it was then stressed would remain in place indefinitely—as well as by U.S. Navy forces in the region.

These clarifications, issued some time after the initial announcement, were warmly welcomed. The logic of role-division as between ROK defense and U.S. deterrence was accepted, and the new emphasis on the indefinite commitment of locally deployed USAF forces, as well as of Navy forces in the region, was greeted with satisfaction. But

Japanese observers pointedly noted that if valid, the logic should apply to Western Europe as well. If the ROK's economy is now well developed, the economies of Western Europe are more so. Why then is the division of roles between local defense and U.S. deterrence to be applied to the ROK but not to Europe? The answer frequently given in the Japanese media was that the real difference was that the American guarantee of the ROK's security was formal, and lacking in sincerity.

In this connection it was duly noted that the peculiar importance and special nontactical role of American ground forces in Europe have long been understood as an essential link in the mechanism of deterrence. Specifically, it was noted that the physical presence of U.S. troops on the ground (i.e. in the path of any hostile action) was generally regarded as a vital credibility requirement of deterrence. The implication of these comparisons with the European theatre was obviously that in the Korean case, the absence of a ground-force link would undermine the credibility of American deterrence in a period of strategic parity.

There was also much comment in Japan on the absence of any strategic evaluation of the consequences of the decision. The possibility that the withdrawal of U.S. ground forces might invite a Soviet initiative (e.g. the sponsorship of a North Korean attack calculated to disrupt U.S.-Chinese relations, by forcing the Chinese to compete with the Russians in extending support to North Korea) was seemingly overlooked. Also ignored was the possible impact on China of a further decline in American military power deployed the region. This would inevitably increase the costs and risks of containment shouldered by the Chinese, thus increasing the incentives to them of moving towards a detente of their own with the Soviet Union. Again, the lack of analysis in this respect was interpreted as a sign that the administration's policy was in fact driven by strictly domestic political considerations, and that it was quite unresponsive to the realities of the region, and even inattentive to the wider American strategic interests at stake in Korea.

Finally there was acute distress over the failure of the administration to exploit its intent diplomatically. In dealing with North Korea, diplomatic action must always have uncertain results, and the possibility of reaching accomodation on any subject must always be low. But the Japanese (and domestic) critics of the withdrawal decision nevertheless felt fully justified in criticizing the administration for its failure to make any attempt to secure a North Korean quid pro quo.

Partly as a result of the sharp Japanese reaction, and partly in response to domestic criticism, the initial decision for a quick and total withdrawal of U.S. ground troops from Korea was soon modified.

Present plans call for a phased withdrawal, with the last brigade of the U.S. 2nd Division remaining in Korea until 1982.

The form of the initial decision and the manner of its announcement have had powerful political repercussions on U.S.-Japanese relations as a whole. In the wake of the fall of South Vietnam, the Korean decision has undermined faith in the *permanence* of the American strategic presence in the region as a whole. The intensity of the Japanese reaction clearly proves that the attempt to treat Japan as an isolated exception in a regional disengagement is ineffectual: a security system is meant to provide a sense of security; if the Japanese do not feel secure, because they do not consider themselves defensible in isolation, then the system simply does not work.

Japan and the North-East Asian/Pacific Balance

Current U.S. assessments of the North-East Asian/Pacific balance tend to dwell on the beneficial consequences of the Russo-Chinese confrontation from the American point of view. It is undoubtedly true that the complex of ideological rivalry—diplomatic competition and territorial disputes—which divides Russians and Chinese has profoundly affected the global balance of power. For example, it is usually estimated that about one-fourth of Soviet ground forces (divisional counts are misleading), and about one-fifth of Soviet tactical air power is deployed against China. China in turn deploys a large proportion of its higher-quality forces in a defensive array against adjacent Soviet forces.

It may therefore seem odd at first sight that the Japanese view the reciprocal build-up of military forces on the Russo-Chinese border with ambivalence. But from a regional point of view the benefits of the situation are certainly dubious: the Russo-Chinese confrontation has in effect diverted Soviet military efforts from Europe to areas which are much closer to Japan. More precisely, as a result of the Soviet build-up on the frontiers of China, a significant share of the net growth in Soviet military power has been deflected from the European theatre to North-East Asia. (Forces on the Sinkiang frontier are of course altogether too remote to affect Japanese security concerns.)

There has been of course a parallel absorption of Chinese military efforts in defensive deployments against the Soviet Union. But seemingly this has not offset the adverse change on the Soviet side of the ledger: China was not, and is not, seen as having significant offshore-intervention capabilities of adequate quality. While Chinese military power is recognized as very great in an abstract sense, it is not perceived as a threat to Japan.

It could be argued of course that were it not for the Soviet threat, the

Chinese would be free to develop their military power in other directions, allocating investments to naval and air intervention forces which would give them a longer strategic reach. However, partly because of entrenched cultural attitudes, and partly because of the nature of the Chinese economy, this latent benefit of the Russo-Chinese confrontation is not deemed significant. The backwardness of Chinese industry is a formidable obstacle to the development of high-quality air power and modern naval forces, and this will not change for a long time. Certainly, in the present state of Chinese industry even the allocation of greatly augmented resources for the procurement of air and naval equipment would not yield much in the way of high-quality intervention capabilities vis-à-vis Japan.

It is in this context that the Japanese view the Soviet-American military balance, and especially the naval balance. As they see it, the trends have been adverse and will remain adverse in the forseeable future. First, much of the decline in the size of the U.S. Navy has occurred in the forces deployed in the Pacific. This is a natural consequence of the disengagement from Southeast Asia (and this disengagement may well have increased the net capabilities available for service in North-East Asia), but the fact remains that in Japanese perceptions there is now a large void which was until recently filled by the presence of American naval forces. Second, Soviet naval forces deployed in the Pacific bases have greatly increased in quality, and are now seen as much more threatening than they were in the past. In part this is due to the greatly expanding activities of these forces, which now routinely embrace the Japanese home islands in their exercises.

While the number of Soviet fleet units, naval bombers, aircraft and air force fighters, and strike-aircraft deployed in or near the Pacific coast has not increased over the last few years, and may even show some slight decline in the near future, the increase in effective capabilities has been very great indeed. For example, diesel-electric submarines are being replaced by nuclear-powered submarines of greatly enhanced capabilities. When a nuclear-powered C-class submarine replaces a 1940s vintage diesel-electric W-class, or two Ws, the increase in the real threat is clearly very great, especially since W-class boats are much more constrained by the geographic disadvantages of the Soviet base complex than the C-class submarines. Similarily, Krivak, Kresta, and Kara warships with modern electronics and missile armaments have been replacing gun-armed destroyers of the 1950s and makeshift missile conversions of the early 1960s. It is now recognized that the principal function of these new ships is to protect the egress and operations of Soviet submarines, rather than to attack directly well-protected surface targets. This does not, however, diminish the threat, since Soviet surface combatants would thus respond to the major Western strategy against Soviet submarines, which calls precisely for the use of U.S. and Japanese submarines to close the choke-points and form barriers in the proximity of Soviet naval bases.

A similar pattern of drastic qualitative improvement is manifest in the deployment of long-range high-performance naval Backfires which have began to complement and replace Soviet naval bombers limited in endurance (Tu-22), or performance (Tu-16) or both (Il-28).

One projection of Soviet naval deployments in the Pacific (from a source inclined to understate rather than to exaggerate the threat)[2] estimates 1985 Soviet naval forces in the area as shown in Table 1.

TABLE 1
Projected 1985 Soviet Naval Forces in the Pacific
and 1975-1985 Ten-Year Change

	1985 Projection	1975-1985 Ten-Year Change
Major Surface Combatants	57	-3
Long-Range High-performance Naval Bombers (Backfires)	45	+45
Submarines		
nuclear, attack (i.e. torpedo)	22	+9
diesel, attack (i.e. torpedo)	21	-9
nuclear, cruise-missile AS	23	+5
diesel, cruise-missile AS	4	-2

In other words, the major upgrading in the quality of surface combatants will only be offset by a very small reduction in numbers; nuclear-powered submarines are expected to replace diesel-electric boats on a more than one-for-one basis, and a respectable force of naval Backfires will appear.

It is true that the quality of American naval forces has been increasing also, and will continue to increase in the future. But the difference in capabilities between a nuclear-powered aircraft carrier with F-14s and a conventional carrier with F-4s, while very great indeed, is not of the same order as the qualitative difference between a P-class/Krivak/Backfire force, and a W-class/Kashin/Tu-16 force. The former was essentially a defensive array; the latter clearly stresses out-of-area capability. Moreover, the qualitative upgrading of American naval forces has taken place concurrently with a drastic decline in the number of fleet units deployed, overall U.S. Navy ship numbers having declined from over 950 to roughy 450 in the ten years 1968-1977. On the Soviet side by contast, the overall numbers have declined only very slightly.

A similar pattern is manifest in the evolution of the Soviet air forces deployed in the Far East. Soviet Long-Range Aviation Backfires are of course expected to be deployed also, adding a fighting capability of immediate concern to Japan. As of now, long-range Soviet bombers and maritime reconnaissance aircraft are active around Japan (including the Tu-95 "Tokyo Express" flights). The MiG-25 which was flown to Japan by a dissident and which seemingly penetrated the air-defense screen without being detected, was of course an interceptor rather than a strike aircraft but this highly publicized episode also served to draw attention to the growth of the general capabilities of the Soviet air force.

The replacement of MiG-17s and Su-7s with Su-19s, of MiG-19s with late-model MiG-21s and MiG-23s, does not represent a mere increase in generic capabilities from the Japanese viewpoint. Unlike their predecessors, these new aircraft have the range needed to operate against the Japanese home islands effectively. Hence an entirely new threat has appeared on the Japanese horizon, a threat usually operationalized in terms of an air-supported invasion of Hokkaido, whose immediate goal would be naval. The gravity with which this specific threat is viewed is demonstrated by the Japanese decision to deploy a high-quality interception force of F-15s supported by EC-121 AEW aircrafts. The USAF deploys four F-4D squadrons at Kadena (Okinawa) and the Marine Corps has an F-4J squadron at Iwakuni (the former are to be reequipped with F-15s). Even with F-15s, however, the 72-aircraft USAF air-defense force in the area does not match the Soviet air forces now deployed in the region.[3]

The Japanese have for a long time assumed that their security was assured by the panoply of U.S. forces in the Pacific. It is only quite recently that they have began to ask what actual forces were contained in the impressive organizational frameworks of the USAF Nineth Corps, the Third MAF in Okinawa, the Fifth Air Force in Japan, and so on. As Japanese media began to convey facts and figures in response, the impact on influential opinion was disconcerting: e.g. a total of just ten USAF tactical-fighter squadrons in the entire vast area embracing the Philippines, Korea, Japan, and Hawaii. In comparison with the threat, this does not appear as a large force, or an adequate one.

The Seventh Fleet in Japanese Perceptions

For both practical and cultural reasons the Japanese are particularly sensitive to the naval forms of armed power. While no doubt aware in an intellectual sense of the rule-setting function of nuclear power balances, and thus of their dominant role in the architecture of general deterrence, naval power looms larger and is of more immediate significance in psychological—and therefore political—terms.

The cultural significance of naval suasion in Japanese perceptions need not be belabored. Japan was opened to foreign access by the sudden appearance of American warships in its waters, and these warships affected Japan not by their conduct of warlike operations but through their mere presence. In this case, political "visibility" reflected a literal visibility: the tactic which won the day was the entry of Perry's fleet in Edo (Tokyo) bay, where the citizens of the de facto capital could see it with their own eyes, thus learning that the Shogunate had truly failed to protect Japan's immunity from external power.[4]

Naturally enough, when the Japanese acquired modern armaments and set out to conquer in turn, naval forces were indispensable to a nation of islands. But the subsequent emphasis on naval power was not entirely natural or altogether appropriate. The major sphere of Japanese action was continental rather than oceanic until 1941: Korea, Manchuria, China. To an important extent, the primacy of naval power over more functional land power was a bureaucratic and social phenomenon: the Navy officer elite attracted the aristocracy much more than the army. This predilection was reinforced by the fact that the Japanese performed much better at sea than on land: the spectacular naval victory over China in 1935 had no counterpart on land; not much glory was to be won in Korea; the laborious and very costly victory over the Russians in Manchuria which culminated at Port Arthur was a much less clean-cut success than the destruction of the Russian fleet in the Straits of Tsushima. Equally in the Pacific war, spectacular naval actions overshadowed the long and inconclusive Japanese campaign in China, and the indifferent performance of army forces in combat with American troops (though of course the Malayan campaign which culminated at Singapore was a brilliant feat of arms).

Since those days the Japanese have learned to fear strategic-air power and nuclear weapons also. But it is not surprising that the formative influence of their naval experiences remains more deeply embedded. Hence the considerable attention focused on their naval balance and its visible manifestations. For example, in the Japan defense *White Papers* published each year over the signature of the director of the Japan Defense Agency (i.e. minister of defense) an entire chapter is regularly devoted to Soviet naval exercises in the seas adjacent to Japan. Recently, this special sensitivity was recognized in the unusual public intervention of the U.S. ambassador to Japan in the U.S. policy debate on naval power. Ambassador Mansfield, the former Majority Leader in the U.S. Senate, and a man who has never been accused of being obsessed with the importance of military power, chose to address a warning to Washington through open media channels in February 1978, his message being that further reductions in the American forces in the Pacific would have very undesirable consequences, given the widespread insecurity already engendered by the seeming impermanence of American power in the region.

The increasingly intrusive presence of the Soviet fleet around Japanese waters has induced a much-increased awareness of the naval balance. More so than ever before, the Japanese now perceive the Seventh Fleet as a counterweight to a hostile naval force which a very wide spectrum of Japanese opinion (most Communists included) sees as a sinister threat. Hence the attention nowadays focused on the actual force-levels of the U.S. Navy in the Pacific, the making of comparisons, and the tabular assessments of the two naval line-ups reproduced in the media.

Another new factor in the situation is the unprecedented sensitivity to the role of U.S. naval forces in protecting the Pacific sea lines of communication, both from Japan to the south into the Indian Ocean and eastwards across the Pacific. The 1973-74 oil crisis brought to the surface what had long been true but ignored: Japan's immediate dependence on foreign oil. Until the 1973-74 crisis, those Japanese who thought of the issue at all tended to focus on Japan's dependence on foreign marketing companies (mostly American) for much of their oil. After the oil crisis, these concerns over market shares were displaced by a wholly more acute and altogether more widespread concern over oil supply as such. The 1973-74 supply crisis originated at source for extraneous political and economic reasons, rather than in transit for military reasons. Nevertheless, an immediate association was made in the popular mind between Japan's newly recognized dependence on foreign oil and the increasingly manifest presence of Soviet naval forces on the sea lines of communication. Thus in the wake of the oil crisis there has been much discussion of Soviet threat scenarios in which tanker traffic destined for Japan would be attacked by Soviet submarines, long-range naval aircraft, or surface combatants anywhere from the Indian Ocean to the vicinity of the Japanese ports of arrival.

The peculiar importance of these scenarios in shaping perceptions of American forces and their importance derives from their exclusive role in the protection of the sea lines of communication beyond Japan's inner security zone. The Japanese Naval Self-Defense Forces are not configured for employment in distant waters; they include no air-capable ships, long-range submarines, or long-range land-based naval strike aircraft, and the support fleet is very limited in capability. The JNSDF is thus primarily of use in defending Japanese territory itself from sea-borne attack, but the direct invasion threat to Japan is *not* salient. Moreover it is generally recognized that Japan is more importantly protected against invasion by both its own and American air power, by land forces (especially in the case of Hokkaido), and—in a more general way—by the U.S. nuclear guarantee. By contrast, there is only the Seventh Fleet to protect the Japanese against the much more salient threat to their sea lines of communications.

For these reasons, it should no longer be expected that force reductions in the Seventh Fleet can take place without attracting public attention and engendering very negative political repercussions. Once of interest only to a handful of specialists, the issue is now very much on the political agenda, and has become a matter of public concern. A security system can fail in deterrence by failing to avert hostile action, and it may fail in war by failing to defeat hostile action (or contain its consequences to levels judged tolerable); but a security system may also fail politically by inducing those whom the system is meant to protect to make alternative arrangements for their own security.

Japanese Perceptions and American Interests

At a time when the United States is spending roughly 5.1 percent of its gross national product on defense,[5] while Japan is spending less than 1 percent, and a time when significant numbers of American workers are losing their jobs to Japanese workers producing for the American market, it may seem inappropriate to focus on a study of American naval power in the Pacific on Japanese concerns, anxieties and distress over some small reduction in the American military effort. In fact it was an awareness of their vulnerability in such comparisons which muted Japanese criticisms of the Korean withdrawal decision. Otherwise they would have been sharper still.

But the disproportion between the American and Japanese defense burden is a strategic given; it undoubtedly raises quasi-ethical issues of equity, and it should undoubtedly become a concern of alliance diplomacy. It is not, however, relevant to a strategic analysis of the role of American naval power in the Pacific, which must instead focus on the relationship between costs and risks of deployment on the one hand, and the imputed effects on American interests on the other.

In evaluating American interests in Japan's security, and American interests in Japanese perceptions of the same, it is common to focus on the economic importance of Japan. It is frequently suggested if only by inference that the sheer size of the Japanese economy, the dynamism of its growth, and its pronounced high-technology content are values in themselves from the American point of view. It is as if a large Japanese GNP were ipso facto a constituent of American welfare. In more refined analyses, attention is focused on the importance of Japan as a market for American goods with no corresponding calculations of the benefits and costs of Japanese exports—both to the American market itself, and to third-party markets where Japanese exports compete with American ones.

Clearly a number of very intricate problems arise in attempting to evaluate the net benefits of a security relationship in the economic sphere. Still more difficult to evaluate are the cultural and scientific dimension of American interests in Japan, and the importance of a free, stable, constructive, and open Japan in the moral economy of Americans at a time when much of the world is dominated by totalitarian closed societies and regressive kleptocracies.

But these analytical difficulties need not be confronted to arrive at a firm judgment. In pondering the consequences of American policies— Far East policies or Navy policies—which increase the incentives to the Japanese of making alternative arrangements for their own security, it suffices to evaluate the costs and risks of "alternative arrangements." While the costs and benefits, economic and otherwise, of the present situation cannot readily be compared to isolate the net benefits of the status quo, it turns out that a conclusive judgment can be reached quite easily when the consequences of alternative arrangements are assessed. The "positive" American interest in the present state of Japan is hard to evaluate in a convincing manner; but the "negative" American interest in the status quo immediately emerges as a highly convincing argument for its preservation. The present state of affairs has its costs and its benefits; the most plausible alternatives are all wholly undesirable.

Although there have recently been some notable defections from this view, the conventional wisdom is still that neither Japan's recent economic travails nor its increasing sense of insecurity can possibly result in any significant reorientation of its foreign-policy attitude which would change the long-standing strategic relationship with the United States. It is argued that the Japanese economy is now inextricably bound with the free-market economies, and especially with that of the United States, which is both the largest export market, and a major source of nonfuel raw materials for Japan. As a result, it is argued that a strategic reorientation would entail a basic transformation of Japan's economy and society, and would far transcend in scope a mere diplomatic realignment.

The argument is cogent but it overlooks the most important peculiarity of Japan's historical development, which is precisely the unique ability of the Japanese to engineer drastic society-wide change in response to changes in the external environment. A century ago, the Meiji restoration (a political revolution) and the subsequent transformation-by-command of Japan's entire society and economy with rapid and drastic change in both customs and structures, were quite deliberate reactions to the forcible entry of American and European naval power. Similarly, in 1945 another society-wide change, initiated by Americans but carried on by Japanese, came in response to military defeat.

Once the possibility of a reorientation is accepted, it became clear that its impact on American strategic interests would be greatly magnified by precisely those economic concomitants which are cited to demonstrate its unfeasibility. What would be the impact on the global military balance if Japanese export capacity were to be retooled to produce military-industrial goods for export to, say, the Soviet Union, instead of consumer goods for the free-market economies? How many army trucks could the Japanese automotive industry produce in lieu of the cars which now compete with the products of Detroit? How many tactical radios, field radars and sensors could be produced by the Japanese electronic industry which now exports color TVs and radar-ovens? And, finally, what military capabilities could Japan itself acquire if it devoted, say, 15 percent of its annual industrial capacity to military production? As of now the Japanese economy is facing a dilemma: additional growth in production cannot be channeled into exports because in many large markets severe protectionist measures are already imminent. On the other hand, without growth, unemployment will relentlessly increase, given the steady progress of labor productivity. Public works are one possible outlet, military production is another; both allow growth and full employment without incremental exports, and the further intensification of the protectionist reactions they cause.

It is premature and unnecessary to predict the specific form that a strategic reorientation of Japan might take, since all the likely possibilities are most unfavorable from the American point of view. A remilitarization of Japan would imply a multiplication of present Japanese capabilities, and the acquisition of nuclear weapons. Much has been made of Japan's "nuclear allergy," but the trend manifest in public opinion polls now reveals that even this deep-seated revulsion is no longer quite as powerful as it was.[6]

A militarily powerful Japan, even if entirely unaggressive, might induce the Chinese to seek an alliance of equals, or else it might drive the Chinese into a renewed entente with Moscow (through the intermediate stage of detente). Both possibilities are fraught with danger from the American point of view. A militarily powerful Japan would certainly frighten the remaining noncommunist states of Southeast Asia into a frantic search for protection unless the American attitude to Asian commitments had by then radically changed. This search would probably lead them to Moscow, whose naval forces and long-range air power could then be complemented by garrison forces stationed in the area.

A reorientation towards China, perhaps with a concurrent move towards strengthened defenses but no full-scale remilitarization, nuclear weapons or intervention capabilities, would seem the most desirable among the changes in the status quo that are plausible. But this alternative would also be extremely undesirable from the American point of view. For one thing, it might provoke a Soviet military move against China and/or Japan designed to weaken China strategically and/or to destroy the industrial base of both countries.

A reorientation towards the Soviet Union, cultural antipathies notwithstanding, would of course transform the balance of power and prejudice in American security very directly.

It is true of course that a full-scale remilitarization of Japan featuring the acquisition of both nuclear weapons and long-range intervention forces would require a long preparatory period, on the order of ten years or more. But the *diplomatic consequences* would be manifest long in advance; indeed the diplomatic conduct of the affected parties might well develop from the start as if the projected Japanese capabilities were already operational and deployed. The well-deserved reputation of the Japanese for energy, skill and cohesion already now engenders deep-seated fears in many parts of Asia even though Japan is by far the least militarized of the industrial states.

For these reasons, the strategic status quo is highly desirable, for all its faults. From the American point of view, a greater Japanese contribution to the joint defense of the region would certainly be desirable. But the outer limit of the military growth that is compatible with American interests is fairly restrictive. In expenditure terms, this ceiling probably does not exceed 4 percent of Japan's current gross national product as opposed to the present 1 percent or less. Anything much beyond this level would undoubtedly entail the danger of triggering an unfavorable diplomatic chain reaction. Within this ceiling, however, Japan could certainly develop fully adequate air defense forces, and also take over the major share of antisubmarine responsibilities in the vast ocean area stretching from Taiwan in the south to the Aleutians in the northeast. The United States would for its part continue to provide naval counterair capabilities (which of course require air-capable ships) as well as the nuclear guarantee.

The present level of Japan's defense effort is much too low from the American point of view. On the other hand, a level corresponding to annual defense budgets of, say, 6 percent of the gross national product, would certainly be much too high, again from the American point of view. Unfortunately there is clearly very little strategic incentive for the Japanese to do more, unless it is very much more. Any moderate increase acceptable to the United States would by definition leave Japan's strategic situation unchanged; the country would remain dependent on the United States for its defense. At the same time, any such increase would still require a proportionate sacrifice in real resources.

On the other hand, there could be a diplomatic incentive for Japan to do more. In fact the present level of Japan's defense expenditure appears to reflect an estimate of the acceptable *minimum* from the American point of view, which probably exceeds the minima set by the need to preserve a core of professional military expertise in the Japan Self-Defense Forces. Since the American-set minimum can obviously be redefined diplomatically, the United States appears to have considerable leverage to influence the level of Japanese defense efforts. In practice, however, this leverage had scarcely been used to rectify the present imbalance in burden sharing.

The View from Peking

China and American Naval Power in the Pacific

The Chinese government has supposedly undergone a fundamental change in the wake of the death of Mao, the eclipse of his protégées and the reemergence of Teng Hsao P'ing. It should be noted, however, that the Chinese attitude to American naval power in the Pacific has remained entirely unaffected by this change in the policies and governance of the People's Republic. Now as before, it is the formal and overt position of Peking that all foreign military bases should be removed from Asia, the Pacific, and indeed worldwide. Now as before, Peking formally insists that all manifestations of out-of-area military power amount to imperialist intrusions, and ought to be removed immediately.

At the same time, in more private—though by no means secret—communications, Peking has followed a quite different line on the desirability of American military power in the Pacific. During the "gang of four" period, Chinese leaders pressed the leaders of Japanese opposition parties to accept the necessity pro.tem. of the U.S.-Japan security treaty; specifically leaders of the Japanese Socialist Party (JSP) (the main opposition grouping) were asked to desist from their long-standing demand for the abolition of the treaty. The JSP and other opposition groups were also told that the Seventh Fleet had an indispensable and highly beneficial role in the area, and that American access to facilities on Japanese territory was no longer to be contested. In effect, this neutralized much of the long-standing political pressure for further reductions in the American use of bases in Japan. Since the fall of the "gang of four," these positions have been reaffirmed categorically by the new leadership.

Similarly, when President Marcos of the Philippines visited Peking in 1975, in the immediate aftermath of the fall of Saigon, the "gang of four" leadership strongly rejected his offer to work towards an expulsion of American naval and USAF forces from their facilities in the Philippines. Members of the opposition in contact with Peking were apparently given the same message. Again, the new Chinese government has explicitly reaffirmed its approval of the strategic status quo in the Philippines.

The strategic logic of the Chinese position has never been spelled out; inference is not however complicated, or ambiguous. It appears that the Chinese perceive three distinct Soviet threats to their security. First, is the nuclear threat. In the early 1970s there was much discussion of the feasibility and possibility of a "surgical" Soviet strike against Chinese nuclear installations and deployed weapons. It was then suggested that for the Soviets it was "now or never": either they would disarm China while Chinese nuclear delivery capabilities remained embryonic, or else they would have to confront the increasing costs and risks that a relentlessly growing Chinese nuclear arsenal would impose upon them.

It is now evident that this conception of the relationship between Chinese capabilities and Soviet options was fundamentally flawed since it implicitly treated the Soviet nuclear arsenal as static, while being focused on the growth of the Chinese. In fact, the Soviet Union is now (1978) in a much better position to carry out a "surgical" strike against Chinese nuclear facilities and deployed weapons than it was in, say, 1970 with or without using nuclear warheads in the process. While the number of aim points constituted by Chinese nuclear facilities and deployed weapons has increased by a factor of two or three (including multiple aim points for semimobile systems, i.e. cave-based IRBMs), the Soviet capabilities have multiplied. In the Su-19 and Backfire the Soviet Union has now acquired highly effective instruments of nonnuclear bombardment, while with the SS-20 it has acquired a ballistic missile of intermediate range (which covers all targets in the PRC) whose accuracy and reliability are very much greater than the SS-4s and SS-5s formerly deployed. As a result, the Soviet "surgical" strike threat to Chinese nuclear capabilities and the associated infrastructure has grown and is still growing.

The second Soviet military threat is conventional military action on land. It is generally believed that a Soviet invasion is effectively deterred by the postinvasion prospect of effective and irreducible guerilla resistance. However, this is only true if an invasion were to extend to the densely populated Han-inhabited parts of China. In Sinkiang, in Inner Mongolia and in the arid western provinces of China proper, the population is either very thin on the ground or non-Chinese, or both. In these peripheral areas Soviet conventional military action could not be effectively opposed by guerilla resistance, and neither could occupation forces be dislodged by such means. (Intervening desert would certainly allow occupation forces to maintain postinvasion security barriers with the east.) In the Sinkiang, Inner Mongolia and arid western provinces the terrain also favors the employment of mechanized forces and tactical air power, given the scant vegetation and generally good visibility. Hence Chinese conventional forces, which are very weak in both mechanized forces and tactical air support, would be fighting at a terrain disadvantage precisely where the Chinese guerilla capability is ineffectual as a deterrent.

China thus faces both a serious "counterforce" threat (nuclear or otherwise) and a land-invasion threat, albeit limited to peripheral zones, where the Soviets might hope to seize critical military and industrial assets (including Lop Nor) but would not control the demographic and political core areas. Paradoxically, however, it seems that it is the Soviet naval threat which is the salient factor in Chinese attitudes towards American power in the Pacific.

The Soviet navy clearly has no significant "projection" capabilities against China. Coastal areas are without exception densely populated and thus even if the considerable coastal-defense capabilities of the Chinese were somehow to fail to defeat, say, the landing of Soviet marine units, local militia would unfailingly suffice to neutralize any possible seaborne invasion. The Soviet navy does not of course have air capabilities for projection of significant size, two or three Kiev VTOL carriers notwithstanding.

Of slightly greater plausibility would be a Soviet naval campaign against Chinese coastal shipping. The Chinese economy is greatly dependent on cabotage; the mass of Chinese MTBS, MGBs and missile-boats could no doubt guard the coastal sea lanes against surface attack but Soviet submarines could still be operated quite freely. But given the composition of Chinese coastal traffic (large numbers of small craft rather than high-value units), the effectiveness of such a Soviet campaign would be doubtful; the damage inflicted on the Chinese economy by any but the most prolonged of campaigns would be moderate. The fact that an antishipping campaign would undoubtedly offer an instrument of coercion distinctly less provocative than either a peripheral invasion, or the (conventional) strategic bombing of Chinese industry does make it a feasible and conceivably useful Soviet policy option but the likelihood that the option would ever be exercised must be rated as very low.

It appears therefore that the demonstrated Chinese concern for the continued operability of the U.S. Seventh Fleet in the Western Pacific is not motivated by the hope that the fleet might deter Soviet naval action against China or her shipping. The remaining possibility is that the Chinese attach great value to the political role of the Seventh Fleet, and that they do so especially because the acquisition of their own naval capabilities will be long delayed by the higher priority given to the expansion and qualitative upgrading of their nuclear arsenal, and second, to the improvement of ground-warfare capabilities, especially antiarmor capabilities. From the Chinese point of view the major Soviet naval threat is political, and the countervailing function of the Seventh Fleet which is of salient value to them is political also.

The insular and peninsular states of Southeast Asia, the Philippines, Indonesia, Singapore, Malaysia, Thailand, and the whole of Indochina, form the natural sphere of Chinese diplomatic influence and the natural markets for Chinese trade. These areas of Asia were long regarded by the Chinese as protectorates, dependencies or client-states, and a long-standing cultural influence is now reinforced by the ideological appeal which the PRC has for certain segments of the local populations, including many of the overseas Chinese in the area. Southeast Asia is accordingly a region where the Chinese presence would quite naturally dominate other foreign influences. But this is not now the case. For one thing the absolute superiority of Japan in all modern technological accomplishments has deprived the Chinese of their historical role as the carriers of technical advancement into the region. China's new policy of technological modernization and industrial development may change this in the very long run, but in the meantime China is trading quite successfully within the area, and Chinese products dominate the market in many kinds of low-cost consumer goods.

But the major deviation from "normality" is strategic rather than economic. Because of the weakness of their own offshore-intervention capabilities, and the growth of Soviet naval power, the Chinese fear that the Soviets will interpose their naval presence between themselves and their natural sphere of interest. This has not happened so far, because the American naval presence has dominated the scene and dwarfed the impact of the Soviet naval activities in the region. If, however, this were to change then the interposition of Soviet naval forces would undoubtedly follow.

If the United States were perceived by the Chinese as an expanding "imperialist" power in the Pacific (as their propaganda sometimes suggests) there would be no reason for them to prefer the American naval presence to the Soviet. But the Chinese evidently do not see things in this light. They frequently assert that the United States is a fading power in the area, and of course the United States is not now perceived by the Chinese as a threat to their security. The Soviet Union by contrast is actively trying to create an anti-Chinese cordon in the area, it is definitely not a fading power in the region, and of course the Soviet Union threatens China elsewhere. For these reasons therefore, the Chinese regard the Seventh Fleet not as a problem but rather as an *interim* solution, a force that can keep the Soviet naval presence under control during a period in which the Chinese cannot do so themselves.

Chinese Perceptions and American Interests

The bureaucratic dichotomy made in U.S. defense planning between the Atlantic/NATO area on the one hand, and the Pacific/East Asian area on the other, obscures the essential linkage between the two: in both cases the principal threat derives from the Soviet Union. Since the gravity of the threat in each area is merely a function of the distribution of Soviet forces, U.S. planning decisions made for one area without reference to the other may be self-defeating. No doubt because to do so would cut across jurisdictional lines and customary distinctions, U.S. defense planning has so far failed to focus on the interaction between the two areas, and continues to consider each separately.

This is unfortunate since the interaction is critical. For example, it is not possible to make a determination of the overall strategic risk in the two areas, and then to redeploy forces from one to the other: the move itself may alter risk levels in *both* areas.

The great difference between the two areas is the fact that in the East Asian/Pacific area the Soviet Union must contend with a formidable third party. While in the case of the Atlantic/NATO area the United States and the NATO allies confront the Soviet Union and its clients essentially unaided, in the East Asian/Pacific area the United States and its Asian allies must contend only with that part of overall Soviet capabilities which is not absorbed in the confrontation with China. It is primarily for this reason that half the deployed U.S. Navy, less than one-eighth of the U.S. Army, two-thirds of the Marine Corps, and a very small part of the U.S. Air Force suffice in the East Asian/Pacific area to balance fully a third of Soviet ground capabilities, roughly one-fifth of the Soviet air force, and over one-third of the Soviet navy. Overall the ratio is clearly very advantageous.

This favorable state of affairs in the region is the key to the European military balance and indeed the global balance. If the Soviet Union could transfer forces from East to West, the already precarious European military balance would become correspondingly more adverse. In a more general sense, Soviet redeployments from East to West would result in a net increase in overall Soviet power, since Soviet forces in the East serve only to defend territory while Soviet forces in the West have a triple function; aside from defense, they assure Soviet political control over the satellites and they are the main instrument of Soviet diplomatic leverage beyond them.

There has been a tendency to regard this desirable state of affairs in East Asia as permanent, as a given of policy rather than one of its concerns. On the argument that the Soviets and Chinese are locked into a historical confrontation, a classic quarrel between adjacent states whose power relation is unstable, it is believed that the present distribution of Soviet military power as between East and West is unalterable also.

This belief equates the underlying confrontation (which may well be organic and permanent) with the actual climate of Sino-Soviet relations (which of course fluctuates); it also assumes as permanent the present distribution of Soviet military power as between East and West. But while a genuine reconciliation and a return to alliance is to be ruled out, there is no reason to believe that the Soviets and Chinese cannot in fact establish a detente of their own as a deliberate and temporary measure of self-interest rather than as a way-station to entente.

It is clear that a Sino-Soviet detente would undermine the entire structure of Western strategy. Over the last two decades, the Soviet Union has been allowed to acquire levels of military power in all dimensions of capability which are only compatible with an adequate security on the assumption that a significant proportion of the increment would be absorbed by the Chinese. This is obvious and well understood. What apparently is often overlooked, is that American policy is an important determinant of the likelihood of a Sino-Soviet detente that would overthrow the assumption.

It is generally agreed that it is the Chinese rather than the Soviets who hold the initiative and who have chosen to bring the confrontation to its present acute state. The Soviet leadership has made repeated attempts to achieve a detente with China, taking the initiative each time only to be rebuffed. (The most public of these attempts took place immediately after Mao's death in September 1976.) If the Chinese have in the past decided to reject a detente with the Soviet Union, they may still change their minds at the next opportunity. American policy cannot of course hope to control Chinese decisions in the matter, but it can clearly affect the balance of incentives and disincentives. Obviously, nothing should be done which would increase the incentives to the Chinese of opting for a detente with Moscow.

This factor has clear implications for American policy in the SALT negotiations, and it has equally direct implications for U.S. decisions on the deployment of naval forces in the Pacific.

It was shown above that the Chinese attach particular value to U.S. naval forces in the Pacific because the latter act as a counterweight to Soviet forces which they cannot now oppose themselves: in the informal long-term program for the increase of Chinese military strength, naval forces are in the last place, and with good reason. Reductions in the strength of American naval forces in the Pacific would therefore add significantly to the costs and risks of the confrontation for the Chinese, thus increasing correspondingly the incentives to a temporary accomodation which would relieve the pressures upon them.

The United States derives an important latent benefit from the deployment of the Seventh Fleet through its impact on Chinese policy, whose repercussions are in turn principally manifest in Europe. A U.S. defense-planning decision to deemphasize naval forces in the Pacific (as part of the overall effort to increase land capabilities for Europe) may thus have the self-defeating result of triggering a political change which would in turn release large Soviet forces for deployment to Europe. Until the 1968-69 Sino-Soviet border crisis only about twenty under-strength Soviet divisions were deployed against China; today, there are forty-two divisions with a large supporting element, and much infrastructure has been built. A Sino-Soviet detente may well release the post-1969 increment in Soviet forces as part of a Sino-Soviet mutual force-reduction agreement. The Soviet Union has repeatedly called for just such an agreement, while the Chinese have refused. American policy should take great care not to do anything which might induce the Chinese to reconsider.

It may seem that the linkage between U.S. naval policies and the European military balance is tenuous, and subject to many secondary uncertainties as well as to the great uncertainty of the motivating factors in Chinese policy in general. But strategy is distinct from a merely administrative process of decision precisely in considering such indirect linkages: when the stakes are very high, prudent strategic conduct must consider linkages far more indirect and tenuous than these. Certainly, the simple bookkeeping that would correlate the distribution of U.S. military power with the magnitude of overall American interests in each given area does not amount to a considered strategic decision.

Notes

1. See for example, George F. Kennan, *The Cloud of Danger: Current Realities of American Foreign Policy* (Boston: Little, Brown, 1977), pp. 107-111.
2. AW & ST March 20, 1978, p. 23. Quoting a PA & E estimate.
3. Reinforcements to the USAF and Marine Corps air contingent could be sent quite easily from the conus, but only if a crisis involved Japan alone and not any other high-priority area. That is unlikely.
4. Earlier, the principal aim of the Shogun's representatives who dealt with Perry had been precisely to dissuade him from sailing to Edo, where the people of the capital and the feudal lords in residence could not see them.
5. The FY 1979 estimate in the DOD Annual Report.
6. See John E. Endicott, *Japan's Nuclear Option: Political, Technical and Strategic Factors* (N.Y.: Praeger, 1975), pp. 91-102.

9

War, Strategy, and Maritime Power

What is a navy in the absence of a maritime strategy? The United States has interests overseas in need of naval protection, and it also depends on much commercial traffic that is maritime. The United States has a large, if diminished, inventory of warships and auxiliaries, as well as sundry ancillary air forces and many shore facilities variously related to naval functions. Just over 500,000 people in uniform operate and administer these ships, aircraft and shore facilities, and another 200,000 operate a complete, self-contained armed force historically associated with amphibious operations, and now still administratively associated with the naval force as such. But the one thing that the United States plainly lacks is a maritime strategy.

What is a navy in the absence of a strategy? It is, in effect, a priesthood. Ships, aircraft and facilities are maintained, as temples are kept clean, repaired and repainted. Fleets are rotated from home bases to overseas deployment areas, and then back again, as liturgical services are performed at set hours, in the days set by the priestly calendar. Routine ceremonies alternate with the consecration of new ships, and with the introduction of new devices, much as new temples are from time to time commissioned to replace those beyond repair, or to augment their number when faith is on the rise, and the harvest gods have been kind. In all priesthoods there are degrees: some priests are confined to the supervision of the lesser sanctuaries of rustic gods; others are deemed elevated enough to officiate at the inner altars where the most powerful gods reside; the analogy with the nuclear guardians in our Navy need not be belabored.

The priests of ancient pagan faiths had many complex tasks, but it was no part of their duty to ask why the sacrifices were made and the prayers chanted. Nor could they question the wisdom of rites or suggest better ways of appeasing the gods. As for those outside the priesthood, they were disqualified to ask questions by their lack of

163

knowledge of the secrets of the faith. And so we ourselves continue with the upkeep of the ships, aircraft and facilities and with their ritual movements—year after year—never asking fundamental questions about our purposes and methods.

Sometimes the peasants rebel and refuse to pay the tithes exacted for the building of replacement temples; sometimes they react at the cost of some new idol made of exotic materials by expensive craftsmen. Then the members of the priesthood unite in their corporate solidarity to evoke all the sinister dangers that will attend the diminution of the number of temples, or the reduction of their magnificence. Sometimes the peasants are successfuly intimidated, and are frightened into paying homage in hard cash; at other times it is the priests who give up, and then they take care not to undermine faith in the temples and idols still in hand, and so they refrain from insisting on the dangers of the gods left unappeased or by temples not built.

What else can a navy do but perform as best it can as a priesthood, if it has no maritime strategy? For only in a strategy may rational ideas be found to inform the choice of sea and air platforms, to provide meaningful guidelines by subsystem designs priorities, and to define the pattern of requisite deployments.

A navy in being is a necessary condition of any maritime strategy but is not a substitute for such. Ever since the defeat of the Imperial Japanese navy the U.S. Navy has had to live without a comprehensive strategy. Now that there is a growing Soviet navy of already impressive proportions, it may seem that a strategy for the U.S. Navy could be found effortlessly, by summing the requirements of defeating the Soviet navy. Unfortunately this easy answer is foreclosed: the Soviet navy itself can find sufficient strategy in the neutralization of American naval power and its alliance adjuncts, but the latter in turn must accomplish positive purposes, and cannot exhaust their function in neutralizing Soviet naval strength.

The United States thus unavoidably needs a positive maritime strategy, i.e., a coherent statement of its own role in the world with a consequent delineation of the maritime requirements of this role. (Maritime rather than merely naval, because to a large extent naval *force* is merely the protective framework for the use of oceans in all its aspects.) The source of the problem is no mystery: we have no maritime strategy because we have no national strategy. But this in turn is no excuse for the failure of the U.S. Navy as a corporate body to formulate a coherent strategy. It merely means that the maritime strategy must be defined in terms of a *presumptive* national strategy, in the hope that the nation will indeed accept the logic of the former, even if it does not fully acknowledge the latter. But this most basic of tasks continues to be evaded. Preoccupied with purely managerial problems, absorbed by the narrow thoughts of bureaucractic role playing, determined to promote these bureaucratic interests through the substrategic devices of systems analysis and all the other numbers games, much more interested in new technology than in the purposeful operation of *all* technologies (and only strategy may confer purpose on mere technicity) our higher naval leadership has not even seriously tried to develop the intellectual structure of a maritime strategy. In some cases there has been the belief that the mere listing of "missions" is a substitute; in others faith has been placed in *posture statements* poised to exploit the latest headlines (e.g., oil in FY 1975 and 1976). It is true that both the internal customs of resource allocation in the Department of Defense, and also our congressional budgetary process demand specifics and are structured to reject rational strategic discourse, as the latter cannot be quantified. The mindless insistence on numbers, even when the absence of strategic context makes the numbers meaningless, is a fact of life. But there is no reason why the Navy cannot develop its own internal strategic discourse and eventually present its own analysis of the nation's maritime needs, even while continuing to feed all the bookkeepers and slide-rule artists with the deceptively precise numbers that they crave. One must hope that the mental corrosion caused by bureaucratic factionalism has not so far developed that the Navy is now in fact incapable of true strategic discourse.

War, Strategy and Maritime Power[1] is not a statement of naval strategy, nor is it a strategical treatise such as would serve directly to guide the formulation of an American naval strategy. It is, however, a most valuable source book that could be of much use to inform the strategic discourse now long overdue. The first group of essays by Bernard Knox, Gordon Turner, Basil Liddell Hart and Norman Gibbs makes a good introduction by addressing the broader problem of war and peace; except for Liddell Hart's notoriously ignorant misapprehension of Clausewitz (he deplores the fellow, plainly never having read him) it is all solid stuff, in a historical vein. The next section has pieces by Herbert Rosinski, Henry Eccles, James Field, and William Reitzel; it focuses more directly on the nature and purposes of strategy itself. Rosinski's contribution amounts to a lucid miniessay that offers what I believe to be the best brief definition of strategy itself, in contradistinction to tactics ("strategy is the comprehensive direction of power; tactics is its immediate application"). Eccles pursues at much greater length and to good purpose the definitional route; neither good nor bad, his contribution is simply basic, and reflects a sustained interest in the fundamentals of strategy that is itself a valuable rarity among us.

The essays by Field and Reitzel, not to be summarized here, are concerned more closely with the specificaly naval aspect, but their focus is on the history of naval strategy rather than on naval strategy *tout court*. What follows after this in the book is a long series of diverse case studies and essays of reappraisal, including Stephen Ambrose on seapower in the two world wars, Martin Blumenson on the continuities and contrasts between the two world wars, and the editor's own essay on the rearmament of Germany, or rather its immediate prelude. Brisk and well written, it is a useful piece of work even for those who have no interest in the past, because it is now easy to see that the issues of 1950-54 are about to reemerge in full force, one way or the other. Readers will want to explore the remaining rich menu of essays on strategic, military, and politicomilitary issues. Necessarily uneven, the average standard is nevertheless high.

Note

1. B. Mitchell Simpson III, ed. *War, Strategy and Maritime Power* (New Brunswick, N.J.: Rutgers University Press, 1977). A collection of articles and essays on strategy and maritime power that have appeared in the *Naval War College Review*, selected and edited by a former editor of the *Review*.

Part III

Styles of Warfare
and Styles of Strategy

10

The American Style of Warfare

National styles differ in war, as they do in the pursuits of peace. Embodied in the tactical orientation of military forces, and revealed by their structures, these national styles reflect not only the material and human attributes of societies but also their collective self-image. That is why the attempt to transplant a national style of warfare into the armed forces of another nation, with a different pattern of strengths, weaknesses and social relations, usually fails. One recalls vividly the failure of the Egyptians to practice Soviet-style armored warfare in 1967, and equally their success with home-grown tactics, at least during the first days of the 1973 war.

To each his own therefore. But even so a fatal dissonance can arise: national styles of warfare embedded as they are in culture and society may retain their domestic authority even while being overtaken by changes in the military environment external to the nation. Particularly dangerous are those changes in the military environment which are subtle and cumulative rather than overt and dramatic. The latter may awaken attention and stimulate a rethinking of military methods and structures which may yet save the situation. But when change is slow and not overtly manifest, things are apt to go on as before, until the sudden and catastrophic discovery of inferiority in war itself.

There is now a real danger that the American style of warfare is being overtaken by precisely this kind of change in the external military environment. Even while the Soviet Union is closing the quality gap in one dimension of military strength after another, and even while our overall military resources are declining relative to those of the Soviet Union, we hold on to our self-image of material superiority. To be sure, the official spokesmen of the services constantly remind us of the growing Soviet advantage in numbers, and the steady improvement in the quality of Soviet weapons and yet the *operational* implications of these facts have not been absorbed. Our national style of warfare remains unchanged: it still presumes a net superiority in matériel, for it is a style based on the methods of attrition rather than maneuver.

We all know what attrition is. It is war in the administrative manner, à la Eisenhower rather than Patton, in which the really important command decisions are in fact logistic decisions. The enemy is treated as a mere inventory of targets, and warfare is a matter of mustering superior resources to destroy his forces by sheer firepower and weight of matériel.

Maneuver, by contrast, is not a familiar operational form in recent American military experience. In fact, in the language of the U.S. Army maneuver is frequently confused with mere movement, or at least offensive movement. Maneuver may well call for movement but it is very much more than that. It can be applied not only in ground combat but in all warfare, and indeed in all things military, even research and development. Maneuver describes "relational" action— that is, action guided by a close study of the enemy and of *his* way of doing things—where the purpose is to muster some localized or specialized strength against the identified points of weakness of an enemy that may have superiority overall.

Maneuver thus depends much more on intelligence (and intellect) than attrition warfare, which can almost be a matter of mere procedure. And maneuver is a higher risk way of fighting. But while the side that has matériel superiority can choose freely between attrition or maneuver, the side whose resources are inferior overall can only prevail by successful maneuver. If an inferior force remains tied by tradition and mental out-look to low-risk/low-payoff attrition methods, it must be defeated. In the cumulative destruction of the forces ranged against one another which characterizes an attrition contest, the inferior force will inevitably be the first to be exhausted.

It is not surprising that maneuver warfare is so unfamiliar to American military men—in whose self-image matériel superiority still looms large—while it is almost instinctive to those who see themselves as inferior in resources, be they Vietnamese or Israelis.

It is by now obvious that the U.S. Army, Navy, and Air Force would no longer enjoy an automatic superiority in matériel if confronted by the forces of the Soviet Union, and yet their structure and methods still implicitly reflect the presumption of a net advantage in resources.

The U.S. Army, for example, has recently promulgated a new manual of tactical doctrine for a major conflict in Europe (FM 100-5). This is a doctrine of pure attrition: Soviet forces are expected to attack in deep columns of armor, and the army means to oppose them by positioning armor and infantry battalions in their path —some pushed forward to act as a "covering force," but the bulk concentrated on the main line of resistance. Advancing Soviet armor is to be defeated by sheer firepower, in sequence: first air attacks well forward of the battle line, then artillery (with precision munitions), then the guns and anti-tank missiles of the yielding "covering force" in a shoot/fall-back/shoot sequence, then the main forces with their own guns, missiles and small arms. Single battalions are to leapfrog one another in a slow withdrawal, to reload with ammunition so that they can resume the orderly administration of firepower. Catch phrases associated with the new doctrine have an industrial flavor: "force-generation," "target servicing," etc. The invading enemy is treated as a mass of individual targets to be destroyed one by one, with the strength of the defense in firepower being ranged against hard armor. No attempt is made to seek out and exploit weaknesses in the *modus operandi* of the enemy or in his array of forces. No thought is given to the possibility of attacking the long flanks that columns of armor must necessarily have. The Army's new doctrine thus continues to presume a new superiority in firepower: U.S. forces are to "mow down" Soviet armor as British imperial infantry once dealt with the Zulu *impis*. The British won, though they were outnumbered as the U.S. Army would be today, but unfortunately the Soviet forces are not Zulus and they will not be outgunned.

A maneuver defense for NATO would be quite another thing. Far from seeking to muster strength against strength in a frontal clash of firepower versus armor, it would rely on attacks against the weak points of the Soviet array. For example, Soviet divisions draw their re-supply from convoys of trucks following in their wake, 1,800 trucks for each tank division and 2,200 for each "motorized rifle" division. Behind each hard wedge of armor there is the soft column of unpro-tected and road-bound trucks. A maneuver alternative to the Army's new doctrine might deploy all-armored and highly agile strike forces which would side-step the oncoming thrust of Soviet armor columns, penetrate through the spaces between the columns, and then advance deeply enough into the enemy's rear so that they could then turn to attack the "soft" traffic of artillery, combatsupport and service units, and supply columns following in the wake of Soviet armor. While U.S. tanks and combat carriers would be formed into these strike forces, the infantry (which is already well equipped with anti-tank missiles) would be placed in the path of the Soviet advance to form resilient and amorphous defense zones. The aim would be to slow down and embed the enemy armor spearheads rather than to destroy them in (costly) combat. In the meantime, the strike forces would be on their way, to advance in parallel to the advancing enemy columns before turning to wade into them. While U.S. battle tanks could no doubt do much better against trucks and artillery carriages than in tank-to-tank combat, the operational goal—as in all genuine maneuver—would not be so much to destroy enemy resources as to dislocate the enemy's scheme of operations. Instead of being faced with an entirely predictable frontal resistance (which they are well organized to defeat) Soviet commanders would be confronted by confused entanglements and sudden emergencies in their own vulnerable rear, as the elusive strike forces shoot up road-bound traffic, only to disappear (when attacked in turn) to come back and attack again somewhere else along the column. Soviet armor spearheads would in some cases run out of supplies while fighting it out in the resilient defense zones; above all, the stream of reinforcement echelons (on which the Soviet method depends) would be drawn away to confront the strike forces in the rear, instead of being fed into the penetrating advance to keep up its momentum.

This is not by any means a fully analyzed idea, and it is of course at the extreme end of the risk/payoff spectrum, but it does illustrate the general principles of maneuver warfare as they apply to all combat— land, sea, or air.

First, one's own high-quality forces must not be expended against those of the enemy; instead, they are to find and attack the weak points in the enemy's array of forces. In the meantime, the enemy's main effort is to be contained (though it cannot be defeated) by a specialized defense, organized from the lower-cost forces.

Second, the key to victory in maneuver is force-disruption rather than destruction. Of course there will be some attrition, but its purpose must be to dislocate the enemy's system of war, rather than to reduce his forces in piecemeal combat. The goal is to force the enemy to abandon his program, rather than just to reduce the forces he has to implement that program.

Finally, maneuver warfare cannot be fought by standard, general-purpose forces shaped by traditional preferences and bureaucratic priorities. Instead, one must deploy forces especially tailored to cope with a specific enemy—that is, forces which are configured to exploit his particular weaknesses, rather than to maximize all-round capabilities. One allows the enemy to dictate one's force structure and tactics; the "organizational initiative" is conceded in order to seize the operational advantage.

An abbreviated air power example illustrates the generality of these rules. Soviet battlefield air-defense systems are now much more formidable in Europe than they were in Arab hands in October 1973, when the Israelis lost almost a quarter of their air force in three days. To do its work, which is to help in the *land* battle, the U.S. Air Force plans to defeat the array of Soviet anti-aircraft guns and missiles by attrition and sheer weight of matériel: special "defense suppression" aircraft are deployed to attack Soviet radars directly, while other special aircraft are to neutralize Soviet radars with electronic counter-measures. In addition, each line aircraft is to carry self-protection electronic devices. In the first few days of a NATO war, when air power would be needed most to give time for the ground forces to deploy, the U.S. Air Force would in fact be busy protecting its own ability to operate at all.

It is interesting to note that others have reacted differently. The Royal Air Force simply cannot afford to fight it out with Soviet air defenses; its plan is to *evade* them rather than defeat them. The RAF has decided to use its aircraft in the immediate rear of the battlefield, to attack Soviet reinforcement echelons rather than the first wave of Soviet forces on the battlefield itself—where defenses are thickest. As some RAF officers see it, the American insistence on taking on the Russians where they are strongest may result in an air force which will be "taking in its own washing" instead of earning its keep. The RAF approach is "relational" maneuver; that of the USAF a form of attrition.

In the case of naval forces, a counter-example can be cited from the opposite side. When Stalin decided to build an oceanic navy as part of the armament program that began in earnest very soon after VE day, his plan reportedly called for a *non*-relational "balanced fleet" on the Anglo-American pattern, with destroyers, cruisers, and aircraft carriers, as well as submarines—the indispensable weapons of the weaker fleet. Had Stalin's successors continued on this path, the Soviet navy would have been a much-inferior imitation of the American and bound to be outclassed in every encounter. But after Stalin's death his naval plans were scrapped and the Russians adopted a relational, "maneuver" approach; they built their own navy specificaly to exploit the weaknesses of the U.S. Navy, instead of trying to imitate its structure. As a result, our surface Navy of carrier task-forces is now confronted by an array of Soviet anti-carrier forces, based on the use of anti-ship missiles carried in submarines, naval aircraft and surface warships. The Soviet navy which this relational scheme has produced cannot do many of the things that the U.S. Navy does so well, but it does have a fair chance of winning a naval war, at least in some circumstances. A non-relational Soviet navy, built to realize the typical naval ideal of a "balanced fleet," would by contrast have guaranteed absolute and total inferiority at sea for the Soviet Union.

Now that the United States has chosen to place itself in a position of military inferiority vis-à-vis the Russians by reverting to the pattern of underspending of the inter-war years, the nonrelational way of doing things with its low-risk/high-cost attrition solutions to every threat is becoming increasingly obsolete. In one area of defense after another there is no third alternative between higher-risk maneuver methods and a guaranteed defeat. In part, the persistence of an obsolete style of warfare is due to an understandable cultural lag. The services are in the position of those remaining ill-informed American tourists who, in Germany or Japan, still offer *sotto voce* to pay their hotel bills in dollars—and expect a discount. But aside from cultural lag there is another source of irrationality, and ironically it is the product of the striving to substitute logic and calculation for military instincts, and bureaucratic goals. Many of the "systems analysis" techniques introduced by McNamara and revived by the present civilian defense chiefs are based on mathematical models which treat warfare as a cumulative exchange of firepower; they are in fact pure attrition models in most cases. Even though the historical record of war shows quite conclusively that superior firepower is often associated with defeat, and that winners more often than not were actually inferior in firepower, these mathematical models continue to be devastatingly influential because they capture all that is conveniently measurable about warfare. Thus bookkeepers may fancy themselves strategists.

Unfortunately, these models miss the essence of warfare, which has little to do with the orderly administration of superior firepower on a passive set of targets. To their great discredit, the uniformed military have chosen to play the bureaucratic game, and now have their own models, suitably rigged. Instead of resisting the pressure to conform, and devoting their intellect to the study of war as it really was in history, and as it may be again on the battlefield, the military waste their talents on studies and models which are based on premises which are false, and which they *know* to be false. Hence the blind lead, and those who could see follow, in order to defeat the mathematics of the civilian "systems analysts" with their own, ever more elaborate computer models. Unforunately, the tactics of bureaucratic conflict in the Pentagon are of no use on the battlefield.

Note

I am indebted to my friend and partner, Dr. Steven L. Canby, for many helpful suggestions.

11

The Operational Level of War

It is a peculiarity of Anglo-Saxon military terminology that it knows of *tactics* (unit, branch, and mixed) and of *theater strategy* as well as of *grand strategy,* but includes no adequate term for the *operational* level of warfare—precisely the level that is most salient in the modern tradition of military thought in continental Europe. The gap has not gone unnoticed, and Basil Liddell-Hart for example attempted to give currency to the term "grand tactics" as a substitute (already by his day the specialized usage of the directly translated term "operational-functioning machine/unit," was too well established to be redeemed.)

The operational level of war, as opposed to the tactical and strategic levels, is or ought to be of greatest concern to the analyst. In theater strategy, political goals and constraints on one hand and available resources on the other determine projected outcomes. At a much lower level, tactics deal with specific techniques. In the operational dimension, by contrast, schemes of warfare such as blitzkrieg or defense in depth evolve or are exploited. Such schemes seek to attain the goals set by theater strategy through suitable combinations of tactics. It is not surprising that the major works of military literature tend to focus on the operational level, as evidenced by the writings of Clausewitz.

What makes this gap in Anglo-Saxon military terminology important for practical purposes is that the absence of the term referring to the operational level reflects an inadvertence towards the whole conception of war associated with it, and this in turn reflects a major eccentricity in the modern Anglo-Saxon experience of war. It is not merely that officers do not *speak* the word but rather that they do not *think* or practice war in operational terms, or do so only in vague or ephemeral ways. The causes of this state of affairs are to be found in the historic circumstances of Anglo-Saxon warfare during this century. In World War I, American troops were only employed late, and then under French direction; their sphere of planning and action was essentially limited to the tactical level. As for the British, who did have to endure the full five years and more of that conflict, they mostly did not transcend their pre-1914 experience, characterized by battalion fights in the colonies.

It was precisely the failure of the British Army to extend its mental horizons that the "English" school of post–World War I military thinkers so greatly deplored, and which it set out to correct. The advocacy of large-unit armored warfare in depth by Fuller, Liddell Hart, etc. was aimed at expanding operations to transcend the tactical battlefield—and was not simply inspired by the need to find employment for the newly invented tank. In other words, their ideas were not tank *driven* but merely tank *using*. The motivating factor was not the attraction of the technology, but rather the powerful urge to escape the bloody stalemate of the tactical battlefields of World War I.

Nor did the radically different character of the World War II suffice to establish the operational level in the conduct, planning, and analysis of Anglo-Saxon warfare. To be sure, there were isolated examples of generalship at the operational level, and indeed very fine examples, but they, and all that they implied, never became organic to the national tradition of warfare. Instead such operational approaches remained the trade secrets and personal attributes to men such as Douglas MacArthur, Patton, and the British General O'Connor, victor of the first North African campaign.

Otherwise, in World War II as in Korea and of necessity in Vietnam, American ground warfare was conducted almost exclusively at the tactical level, and then at the level of theater strategy above that, with almost no operational dimension in between. Thus the theater strategy of 1944 in France (as earlier in Italy) was characterized by the broad-front advance of units which engaged in tactical combat *seriatim*. Above the purely tactical level, the important decisions were primarily of a logistic character. The overall supply dictated the rate of advance, while its distribution would set the vectors of the advancing front. And these were of course the key decisions at the level of theater strategy. Soon after the end of World War II it became fashionable to criticize the broad-front theater strategy pursued after D-Day. But such criticisms overlooked the central fact that the American comparative advantage was in sheer material resources while U.S. (and British) middle-echelon staff and command skills were of a low order. The overly personalized criticism of Eisenhower's strategy that characterized this literature certainly did not result in the popularization of any "operational" concepts of war.

In Korea once again, the predominant pattern of warfare was set by a front-wide advance theater strategy, which practically left no scope for anything more ambitious than tactical actions. The brilliant exception was of course the Inchon landing, but characteristically this experience was assimilated as the virtuoso performance of Douglas MacArthur, instead of being recognized as a particular manifestation of a general phenomenon, i.e., the concerted use of tactical means to achieve operational-level results that are much more than the sum of the (tactical) parts.

Since the Korean War, as before it, American ground forces have continued to absorb new generations of weapons, their mobility in and between theaters has continued to improve, logistic systems have been computerized and much attention has been devoted to the management of resources at all levels. Nevertheless the entire organism continues to function only at the lowest and the highest military levels, while the operational level in between remains undeveloped. This is not due to any lack of military knowledge as such. Rather, it reflects the limitations of an attrition style of war, where there is an exaggerated dependence on firepower as such to the detriment of maneuver and flexibility. In the extreme case of pure attrition, there are only techniques and tactics, and there is no action at all at the operational level. All that remains are routinized techniques of reconnaissance, movement, resupply, etc. to bring firepower-producing forces within range of the most conveniently targetable aggregations of enemy forces and supporting structures. Each set of targets is then to be destroyed by the cumulative effect of firepower, victory being achieved when the proportion destroyed suffices to induce retreat or surrender, or, theoretically, when the full inventory of enemy forces is destroyed.

It is understood of course that in deliberately seeking to engage the largest aggregations of enemy forces, their reciprocal attrition will also have to be absorbed, so that there can be no victory in this style of war without an overall superiority in net attritive capacity. But aside from that, attrition-style warfare has the great attractions of predictability and functional simplicity, since efficiency is identical to effectiveness, and since the whole is (if no more) no less than the sum of the parts. Hence the optimization of *all* military activities in peace as in war, whether research and development, procurement, manpower-acquisition, training, staff work, or command can all be pursued in a systematic fashion—the goal being of course to improve the techniques (target acquisition, force-movement, resupply, etc.) whose combined effect determines the overall efficiency of attritive action. Thus in seeking to enhance overall capabilities, each resource increment can be unfailingly allocated into the right sub-activity, merely by establishing which of them yields the highest marginal return: manpower or equipment, numbers or quality, fire-control or ammunition enhancements, and so on. Under a pure attrition style, all the functions of war and war preparation are therefore governed by a logic analogous to that of microeconomics, and the conduct of warfare at all levels is analogous to the management of a profit-maximizing industrial enterprise. This in turn renders possible the overall management of defense by the use of marginalist analytical techniques, with uncertainties being confined to technical unknowns. Only structural obstacles (e.g. self-serving bureaucracies, or local political pressures) remain to interfere with the pursuit of efficiency.

Thus in the whole complex of war preparation and action, uncertainties are confined to a few irreducibles. Otherwise everything can be routinized on the basis of efficiency-maximizing managerial procedures with the lowly exception of the command of men in direct contact with the enemy, in which nonmanagerial methods of combat leadership remain necessary.

The other main phenomenon of war, which stands in counterpoint to attrition along the spectrum that makes up the overall style of war of nations and armed services is *relational-maneuver*. In the case of relational-maneuver the goal of incapacitating enemy forces or structures—and indeed the whole enemy entity—is pursued in a radically different way. Instead of cumulative destruction, the desired process is systemic disruption—where the "system" may be the whole array of armed forces, some fraction thereof, or indeed technical systems pure and simple.

In general terms, attrition requires that strength be applied against strength. The enemy too must be strong when and where he comes under attack, since a concentration of targets is required to ensure efficiency in the application of effort. By contrast, the starting point of relational-maneuver is precisely the *avoidance* of the enemy's strength, to be followed by the application of some selective strength against a known dimension (physical or psychological) of enemy weakness. While attrition is a quasi-physical process so that fixed proportionalities will govern the relationship between the effort expended and the results achieved, relational-maneuver by contrast does not guarantee any level of results (being capable of failing totally). But neither is it constrained by any proportional ceiling between the effort made and the maximal results that may be achieved. It is because of this nonproportionality that relational-maneuver methods are compulsory for the side weaker in resources, which simply cannot prevail by attrition. But if relational-maneuver methods offer the possibility of much higher payoffs than those of attrition they do so at a correspondingly higher risk of failure. And relational-maneuver solutions are apt to fail catastrophically—unlike attrition solutions which normally fail "gracefully," that is to say gradually.

The vulnerability of relational-maneuver methods to catastrophic failure reflects their dependence on the *precise* application of effort against correctly identified points of weakness. This in turn requires a close understanding of the inner workings of the "system" that is to be disrupted, whether the "system" is, say, a missile, in which case the knowledge needed has an exact technical character, or an entire army, where an understanding of its command ethos and operational propensities will be necessary. Somewhat loosely, one may characterize attrition methods as resource-based and relational-maneuver methods as knowledge-dependent. Both the high potential payoff of the latter, and also their vulnerability to catastrophic failure, derive from this same quality.

Since in any real-life warfare, both pure attrition and pure relational-maneuver are very rare phenomena, what matters is the content of each phenomenon in the overall action, whether that is as narrow as a single tactical episode, or as broad as a national style of warfare or some war preparation activity, such as the development of weapons.

Both attrition and relational-maneuver are still perhaps most familiar in the form of ground warfare. Certainly the most vivid comparison is provided by the contrasting images of the trench battles of World War I on the one hand—symmetrical brute force engagements not far removed from pure attrition—and the great encirclement battles of the 1939-1942 Blitzkrieg period on the other, warfare characterized by low-casualty, high-risk actions. Or to show equal contrast in one national army, in one war and in a single theater of operations, the theater-scale disruptive maneuver of MacArthur's Inchon landing may be compared with the cumulative firepower engagements of General Ridgeway's offensives.

It is to be recognized, however, that both attrition and relational-maneuver are universal phenomena, which pervade all aspects and all forms of war and war preparation. This can be illustrated by a number of direct comparisons, a sample of which are shown in Table 1.

Both attrition and relational-maneuver will be present in all real-life contexts, so that different national (or service) styles of warfare will be distinguished by the proportion of each mode in the overall spectrum, rather than by the theoretical alternatives in pure form.

Having thus suggested the universality of the phenomenon, one may focus on the attrition/maneuver spectrum in ground warfare without fear that relational-maneuver will be confused with more movement, or

TABLE 1
Attrition vs. Relational-Maneuver

	Attrition	Relational-Maneuver
Methods of Target Planning in Strategic Nuclear Warfare	Incapacitate enemy society by destroying high percentage of all industry and all population by the least variable of kill effects (e.g., blast rather than weather-dependent heat).	Incapacitate enemy political-military system by destroying political and military command centers and organizational headquarters; destroy selected critical war-fighting and recovery facilities (e.g., industrial bottlenecks *viz.* straight floorspace). Rely on fine-tuned kill effects.
Deployments of Ground Forces at the Theater Level	Deploy standard-format general-purpose forces to match total computed enemy capabilities. Freely rotate command, staff, and formations between different theaters.	Deploy theater-specialized formations configured especially to exploit the weaknesses of the particular enemy forces in each theater on a long-term basis, with in-theater promotion.
Methods of War Preparation, Research and Development Goals	Develop "best possible" systems to maximize all-round capabilities; hence develop systems *ab initio* to minimize design constraints. Hence long time-lags between generations, and broad changes needed in supporting maintenance structures upon introduction. Thus, only *major* advances can justify development efforts; hence the state of the art must be advanced. Because of long time lags between design and introduction, there will be only a coincidental correspondence between the systems so acquired and the specific configuration of combat needs upon deployment. Engineering priorities lead to revolutionary innovation, from time to time. Final design determined by limits of engineering feasibility and costs.	Examine in detail the relevant enemy forces and weapons. Identify specific limitations and weaknesses. Develop or modify equipment to obtain fine-tuning of capabilities against those forces and weapons. Modify and develop incrementally to maintain a "good fit" as enemy forces also evolve. Since new items are introduced at short intervals, accept design constraints to ensure compatibility (inter-equipment and also with supporting structures). No need to force advances on the state of the art. Create a continuum between in-theater modifications and the central development process.

indeed that attrition itself will be understood only in its narrowest tactical form of a straight exchange of firepower.

One may usefully begin to give concrete definition to the concepts here defined by way of two examples, one well-worn and the other somewhat less familiar, one offensive in strategic orientation and the other defensive, but both examples of *operational* schemes of warfare with a low attrition content: the deep-penetration armor-driven offensive of the classic German Blitzkrieg, and the contemporary Finnish defense-in-depth for Lappland.

The Blitzkrieg Example

The classic German Blitzkrieg of 1939-1942 was an operational scheme designed to exploit the potential of armored fighting vehicles, motor transport and tactical airpower against front-wide linear defenses. Three phases of the overall actions can be distinguished: the initial breakthrough, the penetrations, and the "exploitation."

In the breakthrough stage, axes of passage were opened through the (linear) defenses of the enemy by fairly conventional frontal attacks (and the Germans did so in World War II largely with foot infantry and horse-drawn artillery), but these attacks were focused on enemy forces holding selected, and narrow, segments of the front. The "relational" element of this stage was visible only at the theater level, insofar as soft points could freely be selected for attack (since the immediate areas behind the breakthrough points were of no particular significance in themselves).

The tactical battles fought at the front were not an end in themselves but merely a pre-condition for the next phase. Hence, neither the planning effort nor high value forces were at all focused on this stage. So long as the mobile columns spearheaded by the (scarce) tank forces could gain entry into the depth behind the front, it hardly mattered what happened in the frontal area itself. This in turn allowed the command to choose the break-in points opportunistically, thus already achieving an advantage over a defender whose command remained focused on the tactical battles at the front. The eventual reward of a successful defense against any one breakthrough attempt would be encirclement and capture, once the next phase was executed, through some other breakthrough points.

In the penetration phase, the goal of each mobile column was to advance as fast as possible, eventually to intersect at nodal points deep behind the front, there to cut off the corresponding sections of the frontal defenses.

In a tactical view, the long thin columns of vehicles penetrating through hostile territory were very weak, seemingly highly vulnerable to attacks on their flanks. Tactically, the columns were of course all flank and no "front." But in an operational view, the mobile columns of penetration were very strong, because their whole orientation and their method of warfare gave them a great advantage in tempo and reaction time. Most important, the columns were able to maintain a ceaseless forward movement since they could proceed opportunistically, moving down whatever roads offered least resistance. By contrast, such forces of the defense more capable of organized movement would have to find and intercept the invasion columns, and would thus need to go in specific directions along particular routes, failing in their mission if delayed by the frictions of war or by enemy flank-guard forces that cut across their path.

This strictly mechanical advantage was usually dominated by a command advantage. While the invasion forces did not need detailed instructions—being sufficiently guided by General Mission Orders and by tactical opportunism along the axes of advance—the action of the mobile forces of the defense depended on a command adequately informed of the shape of the unfolding battle. But this was a thing most difficult to achieve; the advance of the invasion columns would in itself generate much more "noise" than signals. Typically, the victims of the Blitzkrieg were left only with the choice of paralysis or potential gross error in "reading" the battle. Flooded with reports of enemy sightings across the entire width of the front and in considerable depth as well, the defending commanders either chose to wait for "the dust to settle" (i.e., paralysis) or else they sent off their mobile forces in chase of the sightings that seemed most credible and whose direction seemed most dangerous. In a situation characterized by the multiplicity of signals thrown out by the high tempo of armor-driven invasion columns, the chances of sorting out the data from the confusion were small indeed.

Moreover, the offense had the advantage of moving vertically across a front organized horizontally, and its advance would therefore cut lines of communication (LOCs), occupy successive nodal points in the road network, and not infrequently overrun command centers, thus further immobilizing the defenders.

These three factors in combination resulted in a net advantage for the offense in the intelligence-decision-action cycle, the decisive factor in all forms of *reciprocal* maneuver.[1] So long as the invasion columns kept up a high tempo of operations, their apparent tactical vulnerability was dominated by their operational advantage since the defender's intercepting and blocking actions would always be one step behind.

A closer look at the process reveals that it was deception that provided security for the main thrusts of penetration, which were hidden in the multiplicity of movements generated by flank-guard columns, side-rails, and "abandoned spurs" in the opportunistic flow of the advance. Actually, deception was inherent to the mode of operations. A successful resistance at any one roadblock would be reported as a victory by the defense and indeed it was, but only at the *tactical* level. Operationally, resistance was made irrelevant, as the invasion columns ultimately by-passed such points.

In the "exploitation" phase, effects purely physical were compounded (and usually dominated) by the psychological effects of the penetrations, and the resulting envelopments. The bulk of the defending forces still holding the front in between the narrow axes of penetration would begin to receive reports of LOCs cut, rear headquarters fallen and famous towns to their rear overrun. At the command level, this precipitated attempts to carry out remedial actions still within the initial conceptual framework of the defense, i.e. attempts to execute "orderly withdrawals" in order to reconstitute a linear front beyond the maximum depth of enemy penetration. But that line of frontal reconstitution receded ever deeper as the invasion columns continued on their way. "Orderly withdrawal" soon acquired the character of a rearward race (with the abandonment of heavy weapons etc.). Since in 1939–1942 large, infantry-heavy forces were trying to race against small armor-mobile forces, the defense, Polish, French, or Soviet, could not win the race. This in turn demoralized the commanders, since even "correct" action was soon shown to be futile. And of course among the troops the abandonment of frontal defenses still intact and often entirely unattacked, news of well-known places behind the front already fallen to the enemy, and finally the actual mechanics of the rearward race (including logistic insufficiency) easily had catastrophic morale effects—not uncommonly leading to the outright disintegration of units.[2]

The exploitation phase culminated in double envelopments with a final stage of annihilation—when the foot infantry, now advancing across the abandoned frontage, finally came to grips with the fragmented forces of the defense trapped within the encirclements.

Since the attrition content of the entire action was low (and indeed almost entirely limited to the breakthrough phase) the decisive level was the *operational*. The power of the Blitzkrieg was not conditioned by the weight of resources employed, and not at all by the firepower of the forces involved; it derived rather from the method of command, from the all-mobile organization of some formations, and from the training, all of which endowed the offense with a systematic advantage in the observation-decision-action cycle. Had the Germans encountered a defender itself superior in the tempo of operations, the tactical weakness of their advancing columns would then have become an operational weakness also, with fatal consequences since: (1) the defending forces on either side of the breakthrough sectors could have "flowed" sideways to close off breakthroughs faster than the enemy could act to keep them open; and (2) the mobile forces of the defense could have intercepted or actually ambushed the invasion columns, thus capitalizing on the inherent tactical weakness of forces which are all flank and no front.[3]

The Elements of the Blitzkrieg Style

Though the following analysis is confined to the operational level,[4] it suffices to illustrate the essential principles involved in the relational-maneuver method of warfare that distinguishes the Blitzkrieg:

The Main Strength of the Enemy Is Avoided as Much as Possible. In the breakthrough phase, avoidance is manifest at the theater level in the fact that only a small fraction of the total frontage is attacked in serious fashion, to break open gaps through which the penetration columns can pass. Hence the overall numerical relationship between the total force of the offense employed in the breakthrough attempt, and the total defending force holding the full frontage, is irrelevant to the outcome. Avoidance is manifest at the operational level in the fact that recognized "strategic" locations are not attacked, the selected points of attempted breakthrough being rather those which happen to be least well-defended (with the proviso that subsequent deep penetrations should be possible from those points). Avoidance is manifest at the tactical level, in the use of "rolling out" tactics to minimize frontal engagements as much as possible. In the penetration phase on the other hand, the salient form of avoidance is tactical: cross-country movement and all the flexibility of opportunism in the detailed routing are exploited to avoid islands of resistance, which are to be by-passed rather than reduced or even encircled.

Deception Is of Central Importance at Every Phase. The breakthrough phase presumes successful deception. While the wedging and "rolling out" attacks are launched against selected narrow segments of the frontage, the bulk of the defensive forces along the unattacked frontage must be prevented from moving towards the intended breakthrough axes by feints and demonstrations all along the front, to mask the real foci of attack. Alternatively, where multiple breakthroughs are attempted, deception can be retroactive insofar as costly persistence is avoided, and whichever breakthroughs are successful are then exploited. Either way, success absolutely requires that the defending command remain in a state of uncertainty. This cannot be achieved by mere secrecy since the maximum period of immunity (even assuming perfect security) could not then extend beyond the outbreak of hostilities. In practice, this elevates the deception plan to full equality with the battle plan; certainly deception planning cannot remain a mere afterthought.

In the penetration phase, deception is inherent in the mode of operation. Unless the advancing columns of penetration move with sufficient speed and directional unpredictability to be masked by confusion, they must be highly vulnerable to attacks on their flanks. While it must be assumed that the progressive advance of the invasion columns will be reported, such "signals" will be masked by the "noise" of the multiplicity of sightings mentioned above. If the signal-to-noise ratio is high, and the defenders can therefore develop a more or less coherent picture of the situation (and do not lose their nerve) then the thin columns of penetration will be as vulnerable operationally as they are tactically.

In the exploitation phase, deception is embodied in the process whereby the columns of penetration cut off and encircle enemy forces that can be much larger than themselves; by then the enemy must be reduced to an incoherent mass (cf. the 1941 battles of encirclement in the Ukraine). The most complete achievement of *systemic disruption* is manifest in the final round-up stage of such battles of encirclement, when the ratio of prisoners of war to captors may be very high indeed. By that stage conventional Order of Battle comparisons between the two sides have lost all meaning. It is obvious that such successes cannot be achieved against an undeceived enemy. Even at a fairly late stage of disintegration, the victim forces could regroup in improvised fashion to defeat the encirclement if they had certain knowledge of a highly favorable force ratio.

The Intangibles Dominate. Momentum dominates other priorities (e.g. firepower capacity and lethality). Even in the breakthrough stage, the "rolling out" must quickly follow the "wedging," for otherwise the forces engaged in the latter become vulnerable to flank attacks. In fact, the breakthrough as a whole must be accomplished rapidly, because otherwise the defense will have the opportunity to redeploy its forces to secure the segments of the frontage under attack—or at least to hold the shoulders firmly. The columns of penetration in turn must pass through as soon as their way is open in order to begin their disruptive process before the defense can react. In fact, the whole operation obviously rests on the ceaseless maintenance of momentum. Organizationally, this implies a very restricted deployment of heavier/slower elements and especially artillery. Even with self-propelled artillery, the need to keep the supply tail light and fast moving will restrict the amount that can usefully be deployed. Tactically, the imperative of momentum will downgrade the importance of accuracy (for lethality) in such firepower as is employed. With the artillery, it is suppressive rather than physically destructive firepower that is wanted. And the same applies to the small-arms firepower of the infantry, the troops being trained for suppressive fire with automatic weapons, rather than for the slow-paced delivery of aimed shots. Technically, this in turn results in a requirement for combat vehicles from which infantry can fire on the move.

It is in the exploitation phase that the importance of force-ratios as such declines to its lowest point, while the importance of sheer momentum is supreme. Accordingly, a progressive thinning down of the advancing columns is preferable to the more deliberate pace that full sustainability across the geographic depth would require. It is not uncommon for the battle to end with the victors depleted and exhausted, their strength reduced to very little at the culminating moment, and in the climactic place of the battle, i.e., where the encirclement pincers close. At that time, in that place, the forces of the offense are quite likely to amount to a congerie of improvised battle groups and assorted sub-units that happen to have reached that far. The implied renunciation of full-force sustainability and formation integrity stands in sharp contrast to the principles of war upheld by attrition-oriented armies (cf. "unity of command" in the U.S. Army).

It is clear that the three operational principles here discussed (avoidance, deception, and the dominance of the intangible momentum) are all interrelated, and indeed their connection is the true essence of all offensive operational methods of warfare that have a high relational-maneuver content. First, the ability to apply "localized or specialized" strengths against the enemy's array of forces implies reciprocally that the enemy's own strength is successfully avoided. That in turn can only be done by deception, since it is only a barrier of ignorance that can prevent the enemy from coming to grips with the attacking forces. Deception in turn can only be sustained if the whole operation has a momentum that exceeds the speed of the intelligence-decision-action cycle of the defending forces. Any one deception scheme must be highly perishable, so that the barrier of ignorance can only be preserved if rapid-paced operations generate deceptive impulses faster than they are exposed as such. It is because of this interrelationship that the decisive level of warfare in the relational-maneuver manner is the operational, that being the lowest level at which avoidance, deception, and the dominance of momentum can be brought together within an integrated scheme of warfare.

The Finnish Example

The Blitzkrieg was offensive strategically, as well as during most tactical phases. It was dependent on the use of armor (even if not at all on any superiority in armor capabilities as such). And of course it was an historical episode repeatable only in special circumstances (e.g. the Sinai fighting of 1967). The Finnish operational method for the defense of the Lappland is by contrast strategically defensive, and tactically defensive also in most respects. It is based on the assumption that no armor at all will be available to the defense. Finally, it is a contemporary scheme theoretically reproducible in a wide variety of circumstances, subject only to availability of expendable space. These dramatic contrasts make the parallelism of operational principles between the Finnish method and the Blitzkrieg all the more persuasive evidence of their universality.

Avoidance of the Enemy's Main Strength

At the level of national strategy, this principle is manifest in the whole conduct of Finnish external policy. Soviet power is deflected by a conciliatory foreign policy. But to set limits on the degree of obedience that Moscow can exact, Finnish policy exploits the "Nordic Balance" in which Soviet pressure on Finland is inhibited by the expectation that it would evoke an increased level of NATO activity in Norway, and a proportionate adjustment in the Swedish alignment towards NATO. It follows at the level of theater strategy that the Finnish contribution to the Nordic Balance by the defense of the invasion corridors to Norway and Sweden is more important than the defense of the major Finnish population centers in the southern part of the country. Hence the most reliably powerful of Soviet capabilities, i.e., to invade the well-roaded South and to bomb Finnish cities, are virtually unopposed. It is the Nordic equilibrium that would deny to the Soviet Union the full strategic advantage of an invasion. Even with Finland conquered, Sweden's adherence to NATO would weaken the overall Baltic position of the Soviet Union. On the other hand, Finnish compliance with Soviet foreign policy desiderata pre-empts intimidation based on the capability of destroying Finish cities. This then leaves Finnish theater strategy with a task that is much more mangeable than either a defense of the South against invasion, or of the cities against air attack—that is resistance against an invasion across the largely uninhabited and mostly roadless Lappland. Even there, the task is not really to *deny* passage to Soviet forces but merely to *delay* them up to a point, and weaken them as much as possible, in order to enhance correspondingly the defensive potential of the Norwegian and Swedish forces in the North.

At the operational level, avoidance is manifest in the form of deployment of defense, and in its mode of action. Far from trying to set up anti-invasion barriers near the border to intercept Soviet invasion columns as soon as they cross, no firm barriers are to be set up at all on the invasion routes leading to Norway and Sweden. Instead, Finnish forces are to operate on either side of the invasion routes, to attack advancing Soviet columns on their flanks after side-stepping their frontal thrusts. Since the Finns can have neither effective air cover from their small air force, nor ground-based anti-aircraft defenses of great value, their protection against air attack must come from dispersal and camouflage. Dispersed Finnish forces arrayed in depth from the Soviet border across the full width of the country are to attack Soviet columns by a variety of hit and run methods, including a multiplicity of raids mounted from whatever cover is available, ambushes where practical, nonpersistent mortar and artillery fires, and so on.

At the tactical level, avoidance is manifest in the fact that the tank and mechanized elements of Soviet invasion columns will not be the main target of Finnish attacks. The major efforts of the defenders will instead be concentrated against supply trucks, artillery trains, and support units—all of which can be attacked effectively without need of ATGMs, or other high-grade anti-armor weapons. In this way, even if Soviet tank and mechanized elements can reach the Norwegian and/or Swedish borders intact, they will do so with their combat-support elements weakened and their supply columns depleted.

Deception

At the operational level, deception is inherent in the structure of the Finnish forces to be deployed in the North. Large and highly visible formations of brigade and divisional size will only be deployed on the southern fringe of the trans-Lappland invasion routes, ostensibly to provide a local defense for the small towns in the area, and chiefly Rovaniemi. The main effort on the other hand, will be mounted by far less visible company sized and smaller units detached from the larger formations, and also by *Sissi* raiding teams (trained by the Frontier Guards) which may operate beyond the Soviet border. The more visible formations of the Finnish deployment will not therefore seem threatening or indeed relevant to the Soviet forces, for which any operation mounted southwards from the invasion routes would be a diversion of effort without strategic meaning.

At the tactical level also, deception will be a necessary part of each combat action. Since Soviet invasion columns will routinely provide flank guards for the "soft" elements following in the van of each armored/mechanized contingent, each Finnish tactical action must be based on two separate elements: a diversionary move, to distract the relevant flank-guard elements, and the attack proper. In a company-sized action for example, one platoon might open fire from a safe distance on the soft elements of a Soviet invasion column to attract the attention of the corresponding flank-guard forces. As soon as the latter move towards the scene, the diversionary platoon will retreat to evade their counterattack while the rest of the Finnish force attacks the now unguarded "soft" elements. Finnish forces will then break off the engagement as soon as possible, to seek safety in dispersal and any cover before regrouping to launch the next action. Similarly, the Finns cannot just mount ambushes against the invasion columns, for any ambush astride the main invasion routes would quickly be defeated by the intervention of Soviet armed helicopter elements and/or artillery fire along with direct attacks. Ambushing actions therefore require that lesser Soviet contingents (and chiefly flank-guard units) be lured into prepared killing grounds by some prior attack against the main columns, followed by a deliberate, enticing retreat. In a battalion-level action for example, a Finnish company may attack the soft elements of a Soviet invasion column, wait until flank-guard detachments arrive on the scene and then retreat from the invasion axis, allowing the Soviet detachments to pursue it until the place of ambush is reached, where the rest of the battalion intervenes.

Dominance of the Intangibles

At the level of theater strategy, the Finnish purpose is to weaken as much as possible the Soviet invasion forces without, however, engaging in costly battles against an enemy so vastly superior in heavy weapons. Hence the imperative of elusiveness. This, incidentally, explains the Finns' lack of interest in the acquisition of modern armor (which the Soviet Union offers to Finland at very reasonable prices) or much modern artillery (Tampella itself produces an excellent 155mm gun-howitzer—mainly for export), or even anti-tank missiles. Only less visible and fully portable weapons (small arms, rocket launchers and light mortars) are compatible with the principle of elusiveness that runs through the theater strategy, the operational method and the tactics. (Even TOW, the principal U.S. anti-tank weapon, presumes motor or helicopter transport; it is not truly man-portable.) Thus the solitary Finnish armored brigade (equipped with Soviet tanks and BRT-50 and BTR-60 combat carriers) is not the nucleus of an armored force eventually to be acquired, nor the tool of a quixotic intent to fight armor with armor, but rather a *training* unit that mimics the potential adversary's war-fighting behavior, used very much in the manner of the USAF's "aggressor" squadrons.

At the tactical level, the small but important *Sissi* elements would fight as outright guerillas with a special emphasis on offensive demolitions while the rest of the Finnish forces would fight as light infantry, using strike/withdraw routines with a heavy emphasis on the tactical use of expedient minefields, to the extent that mines remain available.

Conclusion

So very different in all other respects, the two examples here reviewed share one fundamental thing in common. In both cases, the genesis of the military ideas involved was a recognition that material weakness would ensure the defeat of any symmetrical application of forces. In the German case, the front-piercing Blitzkrieg was the alternative to the *Materialschlacht* on elongated fronts that Germany could not win, if only because blockade would progressively erode the industrial strength of a Germany poorly endowed with raw materials. In the Finnish case, the gross imbalance in military power is such that Finnish forces can only provide a limited war-fighting capacity, in a limited part of the national territory, even when the methods used entail a degree of avoidance which approaches that of outright guerila warfare. (In normal guerila conflict, however, war protracted in time substitutes for depth, whereas in the Finnish case the operational dimension is still geographic depth.)

A sense of material superiority by contrast inspires quite other military ideas and allows other priorities to surface. In the American case historically the goal has been to accelerate the evolution of any conflict with maximal mobilization of the economy for the fastest possible build-up of forces on the one hand, and on the other, the deployment of the largest forces sustainable against the largest concentrations of enemy forces possible, to maximize the overall rate of attrition. A broad-front advance theater strategy directly followed from this, if only because the broader the advance, the greater is the usable transport capacity on the ground. Therefore the larger the force that is deployable, the greater its attritive capacity. At the operational level—a level not at all important in this style of warfare—little more was needed than to coordinate the tactical actions which in turn were simple in nature, consisting mainly of frontal attacks.

The principles of avoidance and deception have not been absent in this style of warfare historically, but they were largely confined to the level of theater strategy. For example, the selection of Normandy for the opening of a second front was of course a most notable example of avoidance and deception. But the selection of northwest France itself contradicted the principle of avoidance—which would have favored other places offering greater outflanking opportunities, e.g. southern France or, better still, the Balkans. At the operational and tactical levels on the other hand, avoidance and deception have been little used, since they stand in direct conflict with the imperative of maximizing the application of force upon the enemy's array. The aim was not to obtain high payoffs at low cost, but rather to obtain *reliable* payoffs on the largest possible scale.

The principle of momentum was manifest only at the highest level of all, the level of grand strategy, whence came insistent pressures for quick results. It was certainly incompatible with broad-front advance theater strategies, which of necessity result in a gradual progression, rather than in rapid penetrations. Nor was momentum compatible with operational methods that amounted to little more than the alignment of tactical actions—or with the tactics. A pattern of schematically predictable frontal attacks would naturally result in gradual step-by-step sequence of forward movement, sustained attack, regroupment, resupply and reinforcement, and then more forward movement, and so on. At both the operational and tactical levels, the goal of maximizing attritive results stands in direct contradiction with the maintenance of momentum; if the integrity of formations must be preserved to maximize the efficiency of firepower production, the speed of the action cannot exceed the rate of forward movement which the artillery and its ammunition supply can sustain. By contrast, in rapid-paced actions, opportunistic routing is *de rigueur* and the breakdown of formations into ad hoc battle groups is virtually routine, so that a progressive decline in the volume of sustainable firepower must be accepted. This is a natural consequence of rapid penetrations in depth, if only because "soft" supply vehicles cannot follow in large numbers until enemy resistance ends.

Of late, as a result of the experiences of Korea and Vietnam, a

"short-war" imperative has emerged as far as Third World involve-
ments are concerned, on the presumption that the contemporary
American political system cannot sustain prolonged conflict. To the
extent that the short-war imperative is accepted, a serious problem
emerges, for it conflicts with a military style that precludes the very
methods that can produce quick results. In this regard, the American
military mindset, still firmly rooted in attrition methods, is *not* congruent with what has become an accepted political imperative. Nevertheless, far from inspiring any structural change, the poor fit between the
political imperatives and the military style of preference has not even
been recognized.

Worse, it also appears that the American military mindset is not
congruent with the European military balance either. In the Central
European theater of NATO, U.S. ground forces are still deployed to
implement pure attrition tactics which presume a net material superiority (or more precisely, a net superiority in firepower production). The
expected enemy, however, is in fact superior in firepower capacity
overall, and would most likely achieve even greater superiorities at the
actual points of contact, where its column thrusts would collide with
the elongated NATO frontage. Current tactics must virtually guarantee
defeat against a materially superior enemy, since strength is to be
applied against strength in a direct attritive exchange.

Given the defensive orientation imposed by the grand strategy of the
NATO alliance, only some relational-maneuver operational method
based on the principles of avoidance (to side-step the major Soviet
thrusts), deception (to mask the defenses), elusiveness (in small scale
counterattacks) and momentum (on the counterstroke) would offer
some hope of victory, although with considerable risks. On the other
hand, it is also true that the politically imposed theater strategy of
Forward Defense precludes the adoption of the only operational
methods that would offer some opportunity to prevail over a materially
more powerful enemy.

Notes

I am greatly indebted to my partner, Steven L. Canby, for many key ideas
developed in this essay.

1. Parade ground infantry drill (right-turn/left-turn) preserves in symbolic
   form what was once a crucial attainment in the maneuver of foot forces.
2. In the German Blitzkrieg of 1939-1942, the particular form of the employment of the *Luftwaffe* had its own powerful morale effects. Since the air-to-
   ground potential was used selectively in great concentration (*viz.* diffuse
   interdiction efforts) troops witnessing the intensive dive bombing of scat-

tered points would form a grossly inflated conception of the power of the *Luftwaffe*.

3. That is indeed what happened in the Golan Heights during the 1973 war from the fourth day of the war, when the Israelis were able to outmaneuver the powerful but slower Syrian tank columns and—in more spectacular fashion—were able to ambush the second Iraqi division sent into combat.

4. The two most important tactics involved in the Blitzkrieg operational method were: at the breakthrough stage, wedging and "rolling out," where concentric attacks by infantry-artillery forces open the way for shallow penetrations by more agile infantry which then widens the initial passage by attacks on the flanks; and, in the penetration phase, the use of light-armor and motorized (including motorcycle) elements as "precursors," to trigger ambushes and to "develop" islands of resistance, so that the tank units can directly by-pass them without delay.

12

Low-Intensity Warfare

Ex cladem, victoria? That old commonplace—From defeat, victory—has it that from the deepest abyss of defeat as from the culminating point of victory, nations start on intersecting paths: the complacent winners to defeat, and the losers who have learned the lessons taught in blood and humiliation, to victory. Actually, history scarcely upholds the commonplace. The defeated may not survive to learn, and of course empires are made by those among the victorious who do not become complacent. Now it seems that this country, already unique in so many other ways, may offer a new precedent to history and a new refutation of the commonplace: the complacent defeated certainly cannot aspire to victory. Three allies and much of our international authority were lost in the Vietnam War as well as much blood and treasure, and yet delusions of adequacy persist. Because of the characteristic ambiguities of that war, the nation, though roundly defeated, has nevertheless been denied the customary benefit of military defeat. Little was therefore learned in the experience, except for two false lessons.

First, the nation acquired its phobia of involvement in the most prevalent form of conflict, and the one form of conflict unlikely to lead to nuclear escalation. The toll that irrational fear has exacted from interests large and small thereby left undefended has continued to grow, since the days of the Angola crisis.

Likewise, it appears that some members of the military profession have come to believe that the armed forces of the United States should not be ordered into war without a prior guarantee of irrevocable public support. They insist on a letter of credit of the sort that is demanded before shipping merchandise to dubious importers from lawless countries. The implicit belief is, of course, that there was no casual link between the *manner* in which the Vietnam War was fought and the increasing aversion of the decently patriotic among the public.

In an alternative formulation, the demand is that the armed forces

195

should only be sent to war if "vital" national interests are at stake. In that case, it may be calculated, public support should endure, no matter how badly the war is fought. Entirely normal and appropriate in the case of Switzerland or San Marino, which have issued no promises to fight in defense of any foreign country, that is of course a bizarre and impossible demand for the United States. Such tranquillity as the world enjoys is in significant degree assured by the defense guarantees which the United States has issued by treaty or otherwise to almost 50 countries around the world. In each case, to honor the promissory note the United States must stand ready to resist aggression even though the interests thereby affirmed can scarcely be deemed "vital," except in the rarest cases. A protective quasi-global empire cannot merely fight when "vital" interests are at stake. That is the privilege of the less ambitious, and in our days neither Britain nor France have claimed exemption. (In 1968 the British Army celebrated its one year of the entire century so far in which no British soldier died in combat.) If, on the other hand, the notion of an imperial obligation to fight for less-than-vital interests is rejected, then in logic one can no longer claim an imperial-sized budget for the armed services, whose quasi-global scope must then be a mere façade, dangerously deceptive to all concerned.

Actually, of course, the lesson in point is quite another: it is an integral part of the duties of the armed forces to sustain public support by a purposeful and decently economical conduct of war operations. Luxuriant bureaucratic excess manifest in lavishly staffed headquarters and absurdly over-elaborate services and—more important—the futile misuse of firepower in huge quantities will, in due course, undermine public support for war even if very important national interests are at risk. Conversely, the elegantly austere conduct of military operations will gain public support even if only minor interests are at stake. Journalists who went to north Borneo to decry anachronism and suspect motives were instead captivated by the romance of elite troops at home in the jungle: after being briefed in rudimentary field headquarters manned by a handful of officers content to sleep in native huts, after going up river in a motorized canoe with three quiet riflemen and a Dayak tracker, even hostile journalists could only write well of them, of the British Army in general, and of the campaign. By contrast, journalists who went to Vietnam favorably disposed (there were a few) could only be antagonized by the experience. They were first confronted by hordes of visibly underemployed officers reduced to clerical duties in sprawling headquarters, and then by scenes of gross tactical excess, the heavyweight fighter-bombers converging to bomb a few flimsy huts, the air cavalry helicopters sweeping a patch of tall grass with a million dollars' worth of ammunition. Some observers could recognize tactical poverty in the very abundance with which the ordnance was used; others could detect the lack of any one clear-cut strategy in the generosity with which each service and branch was granted a role in the war; others still were simply disgusted by the wasteful disproportion between efforts and results.

Public support cannot be demanded up front; it must be earned.

Certainly the large military lessons that Vietnam might have taught have remained unlearned. Notably, the multiservice command system whose apex is formed by the JCS organization and whose sublime Vietnam expression was that great bureaucratic labyrinth known as USMACV, stands totally unreformed.

Still today it ruthlessly subordinates the sharp choices which strategy unfailingly requires to the convenience of bureaucratic harmony between the services and their branches. The "unified" style of military planning and operational control is well suited for a landing and front-opening campaign on the scale of Normandy in June 1944. As soon as the scale is reduced, it results in a grotesque over-elaboration that rapidly becomes dysfunctional.

The other unlearned lesson brings us to our subject: the defense establishment as a whole still operates under the implicit assumption that "low-intensity" warfare is merely a lesser-included case of "real" war. Such "real" war is, of course, an idealized depiction, not based on empirical evidence. Unlike the wars now taking place in El Salvador, Guatemala, Nicaragua, and Peru if not elsewhere in Latin America; in Eritrea, Namibia, and indeed all around South Africa and in the ex-Spanish Sahara, too, in Africa; in Afghanistan, Cambodia, Iraqi Kurdistan, Lebanon, and the Philippines in Asia, the notion of "real" war is not corrupted by the intrusions of complex and greatly varied realities.

Instead, that "real" war for which our weapons are designed, our forces structured, and our officers career-developed (by rapid rotation in any little wars that might be available) lives intact and irrefutable in the pages of our doctrinal manuals, there resting undisturbed because no "real" war has been fought during these last 30 years—and of course one hopes that none will be fought during the next 30 years either. But still the high-intensity, "real" war is obviously the very best of all possible wars for such technically accomplished armed forces as our own, so amply supplied with highly qualified, much-decorated, and well-educated officers whose sophistication would clearly be wasted in the sordid little wars that actually are.

For all its virtues, however, "real war" may not in fact embrace all the equipment requirements, all the operational methods and tactics, and all the organizational formats required for the effective conduct of low-intensity warfare. The latter can be a lesser-included case, but only for armed forces of a particular kind, and not our own.

Attrition, Maneuver, and Low-Intensity Warfare

All armed forces combine elements of attrition on the one hand and relational-maneuver on the other in their overall approach to war; their position in the attrition/maneuver spectrum is manifest in their operational methods, tactics, and organizational arrangements, but especially in their methods of officer education.

The closer they are to the theoretical extreme of pure attrition, the more armed forces tend to be focused on their own internal administration and operations, being correspondingly less responsive to the external environment comprising the enemy, the terrain, and the specific phenomena of any one particular conflict. That of course is the correct orientation for armed forces close to the attrition end of the spectrum. Because victory is to be obtained by administering superior material resources, by their transformation into firepower, and by the application of the latter upon the enemy, armed forces of that kind should concentrate on their own inner workings to maximize process efficiencies all around.

The terrain counts only insofar as it presents obstacles to transportation, deployment, and the efficient application of firepower. As for the enemy, it is merely a set of targets which must be designated, located, and sometimes induced to concentrate. Accordingly, a well-managed armed force of this kind cannot logically be adaptive to the external environment; instead it should strive to develop an optimal set of organizational formats, methods, and tactics which are then to be applied whenever possible with the least modification, because any modification must be suboptimal.

By contrast, the closer they are to the relational-maneuver end of the spectrum, the more armed forces will tend to be outer-regarding. That too is the correct orientation for that kind of armed force. In relational maneuver, victory is to be obtained by identifying the specific weaknesses of the particular enemy and then reconfiguring one's own capabilities to exploit those weaknesses. Therefore the keys to success are first the ability to interpret the external environment in all its aspects, subtle as well as obvious, and then to adapt one's own organizational formats, operational methods, and tactics to suit the requirements of the particular situation.

Accordingly, armed forces with a high relational-maneuver content cannot usually maximize process efficiencies and cannot logically develop optimal organizational formats, methods, and tactics. Instead each must be relational, i.e. reconfigured ad hoc for the theater, the enemy, and the situation. There is, of course, no inherent virtue to either attrition or relational maneuver. Armed forces develop historically to their position on the spectrum, which changes over time, to reflect, inter alia, changes in the perceived balance of military power. The defect of attrition, i.e. its high cost, is balanced by the high risk which is the defect of relational maneuver. In general, it is appropriate for the rich to opt for attrition while the poor who acquire large military ambitions had better also acquire a taste for relational-maneuver, which offers high payoffs of low material cost in exchange for corresponding risks. The trouble begins, and the equality between the two approaches to war ends, precisely in the case of low-intensity warfare. Then we find that between armed forces of equal competence, the closer they stand to the relational-maneuver end of the spectrum, the greater will be their effectiveness.

That result follows inexorably by definition: in the degree that intensity declines, the relevance of attrition must decline also, simply because the targets become less and less defined, and more and more dispersed. Yet more important, the dominant phenomena of war become more and more insubstantial and untargetable: not even the most accurate of our precision-guided munitions can be aimed at an atmosphere of terror or at a climate of subversion. The obdurate pursuit of attrition efficiencies in a low-intensity conflict can only be futile. And the greater the volume of the "throughputs" that are processed to generate firepower, the more the results are likely to be counterproductive by antagonizing the local population, which must suffer collateral damage, by demoralizing the armed forces themselves, whose members must be aware of the futility, and by arousing opposition within the nation at home, for even the firmly patriotic cannot but react adversely to a great and costly disproportion between vast efforts and dubious results.

Without attempting to cite an exhaustive record, it is by contrast interesting to note the success of the prototypical relational-maneuver armies when they tried their hand at low-intensity operations. Now that the mists of wartime propaganda, and of the patriotic self-delusion of the occupied nations, have both been dissipated by serious historical research, the success of German counterguerrilla operations in Greece, Italy, Yugoslavia, and France had been duly recognized. As usual with the German Army, relational organizational formats and tailor-made operational methods played a large role in these successes. Similarly, the total absence of a documentary record should not cause us to overlook the outstanding success of the Israeli Army in virtually extinguishing both guerrilla and terrorist activities in the Gaza Strip and the West Bank. Again, novel operational methods tailored specifically to local peculiarities played a large role in the outcome, as did a great variety of specially designed relational equipment.

How Not to Do It

In theory, armed forces endowed with competent leaders should adapt to diverse circumstances regardless of their original orientation. But in practice, as noted, the greater their attrition content, the more will armed forces tend to be inner-regarding, eventually reaching a point where they scarcely extend diplomatic recognition to the actual phenomena of any one particular conflict, especially if those phenomena are complex, ill-defined, and ambiguous—as is usually the case in low-intensity conflicts. When, in additon, the armed forces also happen to have an exceedingly complex internal structure greatly over-officered, pervasively over-administered, and minutely regulated by inter-bureaucratic compacts between services and branches, all the rigidities that ensue will further inhibit adaptation. For one thing, the internal coordination of the diverse forces (and the accompanying office politics) will absorb much of the energy of staffs and commanders. Beyond that there is an even greater obstacle: in the nature of things, any sharply cut adaptive response is likely to attack the delicate fabric of bureaucratic harmony.

It was only logical, therefore, that in Vietnam USMACV should have developed into an impressively large headquarters devoted to the "equitable sharing" of the war among the services and their branches. No organization so complex on the inside could possibly be responsive to the quite varied and often exotic phenomena on the outside. Instead, under its loose and most generous administration, each element was allowed to perform in its own preferred style, often to produce firepower in huge amounts in spite of the great scarcity of conveniently targetable enemies.

Because the system has not been reformed to produce our own version of a non-service, non-branch General Staff, we can expect no better result in the future. Let the United States go to war, virtually any war, and we would again see the Air Force's Tactical Air Command bombing away, and the Strategic Air Command too, most probably; if there is a coastline anywhere near, the Navy will claim two shares, one for its own tactical air and another for the big guns of its gloriously reactivated battleships; none would dare to deny the Marine Corps its own slice of the territory, to be entered over the beach if physically possible, even if ports happen to be most convenient.

Nor can the Army be expected to harm its own internal conviviality by failing to provide fair shares for all, armor even in the jungle, artillery even if the enemy hardly gathers, and so on. After all, a "unified" command and bureaucratized services can only reproduce their own image, and if the enemy refuses to cooperate by playing his assigned role in everyone's conception of a "real" war, the discourtesy will simply be ignored.

Just recently, for example, it was decided to have an exercise in Central America. Aside from both the Second and Third Fleets, legitimately present, room was found to employ both the Seabees and the Army Engineers for a minor bit of well-drilling and such; both the Marines and the Coast Guard were deemed essential to train a few Hondurans in the handling of a few small boats; of course the Marines figured again as a force which must arrive on the scene by amphibious landing; and finally, to train another few Hondurans in counterguerrilla operations, it was deemed essential to employ the Army's Special Forces and the Navy SEALs and a Special Operations detachment of the Air Force.

Undoubtedly the Hondurans should be grateful for such a varied generosity; one need only think of all the pleasant hours that their officers and men will pass in future years as they compare and contrast all those different procedures, diverse jargons, and contrasting doctrines that they saw applied to the same few tasks. There can be no greater affirmation of our national commitment to pluralism.

The "unified" method of military action yields for us all the economies typical of multinational alliances and also their typical degree of strategic coherence—without, however, supplying foreigners to do some of the dying. But the "unified" style does have a surpassing bureaucratic virtue: it can justify large overheads for small operations. With a sufficient degree of organizational fragmentation, the labor of coordination can become wonderfully complicated even if only minute forces are involved. Thus notoriously overstaffed headquarters are allowed, if only briefly, to experience the joys of full employment.[1]

Fight Separate and Win

In theory, competent military leaders should be able to adjust the practice of their armed forces to achieve an optimal position on the attrition/relational-maneuver spectrum, according to the relevant military force balances and the situation at hand. In practice, however, it is history (as fossilized by tradition) and also the collective self-image of the armed forces and the nation itself that determine the composition of the attrition/relational-maneuver mix. If, therefore, armed forces with a high attrition content must engage in low-intensity warfare, the best option is to create a separate force for the purpose.

Because the influences to be overcome are so pervasive, the more the dedicated low-intensity force is separate in every way from the rest of the armed forces, the greater its chances of success. In practice, when the attrition content of the armed forces is extremely high, it is not merely specialized units that are needed but rather a separate branch so autonomous that it begins to resemble a separate service. It certainly needs its own officer corps trained for the task *ab initio* and placed in a separate career track.

Every instinct of bureaucratic efficiency is against that solution. But for armed forces inherently ill-suited for the conduct of low-intensity operations, but which may be highly effective in other roles, the separatist solution is the only alternative to failure, or else severe deformations.

Certainly the attempt to change over to a relational-maneuver style merely to engage in low-intensity war must be disruptive and potentially dangerous. One can easily contemplate the consequences that would have ensued if the United States had in fact won the Vietnam War in relational style, by converting its Army into an Asian constabulary.

On the other hand, it is simply unprofessional to try to fight a low-intensity war with forces structured and built for the opposite requirement. Consider four profound differences:

- Armed forces with a high attrition content are supposed to optimize standard operating procedures for worldwide application, because for them all wars are the same. Low-intensity wars, however, are all different, and each requires an ad hoc set of standard operating procedures. It follows that a primary task for the officers of the dedicated body is to develop one-place, one-time adaptive doctrines and methods.
- Armed forces with a high attrition content must treat all their personnel as interchangeable parts to maintain their efficiency. Low-intensity wars, on the other hand, usually require the persistent application of one-place, one-time expertise, embodied in specific individuals with unique attributes. Thus the normal practices of rotation cannot apply.

- Armed forces with a high attrition content operate within an arena of military action demarcated by externally-set political guides. Low-intensity wars, however, are made of political phenomena with a martial aspect. It follows that the senior officers of the dedicated body should have the particular aptitudes needed for the successful manipulation of the political variables. In low-intensity wars victory is normally obtained by altering the political variables to the point where the enemy becomes ineffectual, and not by actually defeating enemies in battle.
- Armed forces with a high attrition content must accord a dominant priority to logistics first of all, and then to the deployment, upkeep, and utilization of the best-available means of firepower. Low-intensity wars cannot, by definition, be won by the efficient application of firepower. It follows that the officers of the dedicated body do not need the skills and aptitudes required for the management of large-scale organizations and the efficient operation of advanced equipment. On the other hand, they do need the ability to insert themselves into a foreign cultural milieu and to train and then lead local forces or native auxiliaries, who will almost always be equipped only with the simplest weapons.

The sublime irony is, of course, that the United States already has such a dedicated body, although not sufficiently autonomous to offer a separate career track. By nature "relational," by nature adaptive, the Special Forces should be exactly what we need. Their very existence is an implied recognition that low-intensity war is not a lesser-included case; this contradicts the dominant orientation. Hence the existence of the Special Forces has always been precarious.

At present, the Special Forces are very weak bureaucratically because they are merely marginal when they should instead be autonomous and yet also accepted as an important part of the Army. From this all the other evils derive, including the Special Forces' difficulties in attracting the more ambitious among the officer cadre, and the observed propensity of the "unified" commands and the JCS to push Special Forces aside as soon as a conflict begins to look role-enhancing to the bigger boys. One possible solution is to solve the problem by an act of political intervention—more sustained and effective than President Kennedy's initiative. Another and far superior solution is to create a broader framework in which Special Forces would naturally fit and from which it could draw support: a light infantry branch whose several divisions—much needed in any case—would have a pronounced relational-maneuver orientation and which would be outer-regarding by nature.

One consequence of the Special Forces' bureaucratic weakness, seemingly quite petty but in fact revealing and by no means unimportant, is vividly manifest in El Salvador. It is a typical assumption of inner-regarding armed forces that their particular equipment preferences have universal validity. As a result, it is assumed that by appropriate selection from the standard inventory any particular war requirement can be met.

More remarkably still, it is implicitly believed that the equipment developed to suit the needs and possibilities of the richest armed forces of the world will also fit the needs of the motley forces which are invariably our allies in low-intensity wars. For example, the U.S. Army and Marine Corps both happen to favor the lightest, cheapest, and least capable of the automatic rifles on the world market. That is an understandable preference for armed forces which actually plan to fight their "real" wars by artillery and airpower. Under the inner-regarding practice it is assumed as a matter of course that the same rifle will also be suitable for the army of El Salvador, for whom rifles and machine guns provide virtually all the available firepower. Our late allies in Indochina were given M-16s, and now the troops of El Salvador receive the same flimsy and unsoldierly rifle, with the same millimetric tolerances that require standards of cleanliness unknown to peasants. Acres of computer printouts may prove the excellence of the weapon, but one should not expect high self-confidence from soldiers who are sent into action carrying a weapon that feels like a large toothbrush. But then, of course, there is no mathematical designation for "feel," and no system preoccupied with "real" war can be expected to pay attention to such petty things as mere rifles.

Certainly if the Special Forces had anywhere near the appropriate degree of autonomy, they would long ago have ensured the production of a sturdy steel and wood "military assistance" automatic rifle—a U.S. AK-47, similar to the Israeli AK-47 which has been embellished into the Galil. These would, of course, be demonstrably inferior to the M-16 by any respectable operational research (the Galil is downright absurd because of its weight), but you could bet your paycheck against an old copy of FM 100-5 that every self-respecting soldier in the Army would seek to have the sturdier, better-feeling weapon.

Another obvious requirement vividly manifest in El Salvador is the production of a "military assistance" machine-gun more forgiving of human frailties than the M-60. That, too, is a perfectly good weapon, of course, but rudimentary armies are better off with a magazine-fed light machine-gun that is more difficult to jam.

Far more important is the strategic autonomy that would result from institutional autonomy. Low-intensity wars should belong to the Special Forces unambiguously and fully, with other service components coming in as needed under Special Forces direction, to be the servants and not the masters.

In the terms of the art, low-intensity wars would come under "specified" commands set up for the purpose and headed by Special Forces officers. Then, one hopes, we would no longer see even the smallest military assistance groups shared out between the different services and we would no longer see the constant renewal of inexperience by the senseless enforcement of the principal of rotation even in cases where unique expertise vital for continuity is thereby dissipated.

It was not because of any deep-seated cultural defect in the nation as a whole, nor because of a lack of dedication, talent, or expertise in the armed forces that the Vietnam War was lost in the sequence of gross excess, public opposition, imposed withdrawal, and final abandonment. It was rather the uniquely inappropriate organizational structure of multi-service armed forces structurally dedicated to the conduct of "real" war in the attrition style that condemned so many good men to perform so very badly. It is imperative now to achieve the drastic organizational remedy that will liberate the abilities and patriotic devotion so amply present among officers and men, to obtain their fruits for the nation.

Notes

This article is an edited version of a paper presented by the author at the conference "Defense Planning for the 1990s and the Changing International Environment" held at Fort McNair, Washington, D.C., on 7-8 October 1983, cosponsored by National Defense University and the Assistant Secretary of Defense for International Security Affairs.

1. The headquarters and service units sent into Honduras in conjunction with the exercise attained impressive dimensions: 1500 were assigned to support 3500 (*The Washington Post*, 24 August, p. A22). That is the sort of ratio that inspires all the ill-concealed ridicule of the militarily competent among our allies.

13

Acrobats, Architects, and the Strategists of the Nuclear Age

Review of *The Wizards of Armageddon: Strategists of the Nuclear Age,*
by Fred Kaplan

This account of the evolution of American strategic thought in the nuclear age begins in a promising fashion, with the first attempts to understand how the atomic bomb could be used, or rather kept unused, to keep the peace. Among the first to study the matter were the classical economist Jacob Viner, the equally classical political philosopher Arnold Wolfers, and the young Bernard Brodie, already well known as a naval historian. Brodie soon transcended his elders because, unlike them, he was deeply interested in the technical aspect of the new weapons; he understood, for example, that the fission bomb of those days could destroy cities far more easily than armies in the field, especially if they were tank armies.

The quest for new ideas was soon encouraged and supported by the most progressive military figures of the day, the generals of the Army Air Corps and emergent Air Force, notably Henry "Hap" Arnold and Curtis LeMay. They were responsible for the creation of Rand, the first of the nonprofit government-funded "think tanks." Rand was launched as an Air Force project for the explicit purpose of providing a permanent peacetime home for the sort of civilian scientists and "operational analysts" who had been found so useful in the war just ended. To attract talented minds who would not want to remain in the peacetime civil service, Rand was designed to combine the virtues of a non-bureaucratic research institute with the long-term funding and secret information which only the government could provide.

It was in those earliest years of the nuclear age that the ancient idea of deterrence (already explicit and indeed prominent in Roman times) was given its modern role, at the very center of strategy. Here Kaplan has good material in the doings and sayings of interesting and talented men, and he uses it well. Yet there is also a curious omission: Kaplan does not feel it necessary to describe, even in a single phrase, what it was that the American deterrent was to deter, namely Stalin's Soviet Union which was even then being sealed off from the outside world more fully than ever before, which was subjecting Eastern Europe to novel political brutalities, and which was probing for expansion most energetically from Norway to Iran.

Nor does Kaplan think it interesting to ask why it was that seemingly not a thought was given to the possibility of extracting concessions from the Soviet Union by the threat of using the bomb, then still an American monopoly. As a matter of fact, Kaplan does not record even the modest defensive benefit the bomb did bring merely by its existence: the tacit rules of conduct that allowed the United States to protect key interests in spite of the superior might of the Soviet army, as most notably in the first Berlin crisis. It was only because of the invisible protection of the bomb that the Soviet land blockade could be circumvented by entirely vulnerable American transport aircraft which landed safely in West Berlin right over Soviet anti-aircraft guns.

It is obvious enough why the atomic monopoly served only passively to protect American interests from further aggressions, and not to force Stalin back from Central Europe, as Czar Alexander I had been forced to withdraw after Napoleon's defeat. The reason was that an American society which has so recently refused to fight until forced to do so, would not now even contemplate the deliberate use of the bomb for a diplomacy of coercion. But why did military men, sometimes depicted as downright bloodthirsty, not even try to argue the merits of an American *pax atomica?* Would Stalin have shown similar restraint had the atomic monopoly been his?

This most fundamental difference between the Soviet Union and the United States is the central truth of the strategic contention between the two: the United States is most reluctant to use such power as it has, whereas in the Soviet system there is compulsion, partly ideological and partly bureaucratic, to use power to the very limit. But it is precisely this truth that Kaplan avoids, for, as one soon discovers, what he is trying to do is to show American intellectuals and military men collaborating mostly to promote the "arms race." So too he avoids all reference to the purely defensive benefits which the bomb did bring in the 1945-49 period of U.S. monopoly, because it is now *de rigueur* to insist not merely on the dangers of nuclear weapons, but also on their complete uselessness for the purposes of statecraft.

But the trouble with Kaplan's book does not become fully apparent until one reaches page 74 and the chapter on the role of the Rand men in the decision to develop the thermonuclear "superbomb." There was of course a fierce controversy, supposedly scientific but actually political: in a malpractice that was to become habitual, scientists who actually opposed the weapon for political reasons usurped their scientific authority to argue that such a bomb was not technically feasible (the current rendition of this practice is aimed at President Reagan's new strategic defense initiatives). Kaplan makes much of Edward Teller's personality in explaining why it was that the H-bomb was developed at all. Teller, he writes, was "an almost fanatical anticommunist with a particular loathing for the Soviet Union." In contrast to a wise and almost saintly J. Robert Oppenheimer, who did not want to build the H-bomb, Teller is described as "furious, driven," and obsessive.

Kaplan, judging by his citations, went to great lengths to uncover obscure unpublished sources and to extract documents from the Pentagon by using the Freedom of Information Act. All this activity must have interfered with the plain reading of widely available printed books. Had he consulted Dean Acheson's *Present at the Creation*, he would have discovered that the H-bomb decision was not the product of Teller's individual character and supposed political beliefs but rather followed from lawyerly deliberations on two key questions: could it be built? and could the Soviet Union build one soon? Acheson and his aides (chiefly Paul Nitze) examined the evidence and decided that the answer to both questions was yes. That made the decision to build the weapon inevitable.

Kaplan ends the H-bomb chapter with a little anecdote calculated to depict Teller as an unfeeling monster. When the H-bomb device, code-named "Mike," duly exploded in the first test, Teller "in a fit of joy" sent a telegram to the director of the Los Alamos laboratory: "It's a boy." It is only some thirty pages later, and then by tangential reference in a chapter devoted to Albert Wohlstetter and his pervasively influential basing study (which inaugurated the centrality of force vulnerability in American strategic-nuclear thought), that Kaplan causally mentions the fact that the Soviet H-bomb was detonated in August 1953, a mere eleven months after the American test. And it is still some pages after that, in discussing Rand's role in instigating the development of the ICBM, that Kaplan refers, again tangentially, to the fact that the Soviet H-bomb employed lithium. This was no small matter: it meant that the Soviet H-bomb was in fact a *bomb*, a practical weapon, deliverable by an aircraft or even a missile. "Mike," by contrast, had only been an experimental device made huge and nondeliverable by a refrigeration apparatus. Now Rand's influence and Teller's supposed obsessions seem not so important after all. Seriously delayed by Oppenheimer's opposition, the American decision to build the H-bomb had come just in time.

Obviulsy, the Soviet H-bomb could not have been the product of the "action-reaction" phenomenon that supposedly drives the arms race. The Soviet Union's test of a bomb in August 1953 suggests a decision made by 1948 or even earlier, certainly long before Dean Acheson arbitrated between the contending scientists. Kaplan is writing about Americans, and the evolution of American thinking about the strategic questions of the nuclear age, but given his depiction of Teller, and his description of the Rand crowd as "collaborators" in the awful deed of building the H-bomb, we should ask ourselves how it was that the Soviet Union, quite Randless, started to build the H-bomb even earlier. Who was the Soviet Teller? Was he perhaps Hungarian too? Did he also have bushy eyebrows, and was he "an almost fanatical anti-capitalist with a particular loathing for the United States"?

Nor does Kaplan pause to consider the profound and sinister implications of the fact that the Soviet H-bomb decision was made so early. For the circumstances of the Soviet Union would easily have justified a long delay in acquiring the weapon. A wealthy United States, far more advanced scientifically but very weak in military power of the conventional sort (in 1950 American forces could scarcely cope with the North Koreans), could easily obtain the bomb and also had a great need of it, then as now to contain the far superior armies of the Soviet Union. For the latter, by contrast, the H-bomb was a far greater sacrifice of far smaller utility. Nevertheless, in a Soviet Union where the war's destruction had compounded colossal mismanagement to bring back a harsh poverty of miserable overcrowding, ragged clothes, and sheer hunger, the decision to build the H-bomb was made even earlier than in the United States.

That uncovers the second central truth of the strategic contention between the two sides so misleadingly equated in the very term "super powers," and it too is a truth ignored by Kaplan. It is that there is no symmetry of effort between Americans who so reluctantly use a small fraction of their vast wealth to acquire military power and a Soviet regime which imposes enormous sacrifices on its own people and colonizes others in order to increase its military power.

With the entire strategic reality of the Soviet Union thus excluded from the picture, Kaplan is off and running. In writing of Paul Nitze and his role in NSC-68 (the plan for conventional rearmament prompted by the Soviet A-bomb in 1949 but implemented only in part, and then only after the North Korean invasion); in describing the Gaither committee and the ballistic-missile decisions it advocated (in the wake of the Sputnik shock); and in discussing the Rand thinkers, Kaplan combines a lucid exposition of ideas and procedures with the insistent suggestion that Nitze and the rest were afflicted by almost hysterical fears or, worse, that they consciously manipulated the evidence to arouse groundless fears.

With not a word about the Gulag, Budapest 1956, or Khrushchev's threats of nuclear bombardment against Paris and London in that same year (on the occasion of the Suez invasion), and above all with nothing whatever said about the Soviet army, we are invited to contemplate arrogant and yet weak American minds relentlessly threatening the Soviet Union with unnecessary arms.

In the same vein, when Kaplan writes of the Strategic Air Command, Curtis LeMay, and the persistence of a rigid, all-out bombing plan (which anticipated and outlasted the policy of "massive retaliation"), the text is again well informed and informative about all the detailed aspects and yet simultaneously tendentious in the extreme. Now Kaplan offers us the spectacle of men cold-bloodedly planning the destruction of Soviet cities but he gives no serious consideration to the strategic predicament which was the source of it all: to protect Western Europe from intimidation if not attack, the United States had acquired all the attributes of a European Great Power except the most important, a great army.

The thinking and operational studies of Bernard Brodie, Herbert Goldhammer, Herman Kahn, William Kaufmann, Nathan Leites, Andrew Marshall, Henry Rowen, Albert Wohlstetter, and other Rand strategists; the planning and force-building of the military establishment and notably the Strategic Air Command; and the policy advocacy of Paul Nitze, Robert Sprague (the true author of the influential report named for Rowan H. Gaither), and a good many others since, including the Committee on the Present Danger, have all been prompted and shaped by the great asymmetry between the Soviet Union's powerful ground forces and their changing but always inadequate Western counterparts. To keep a tolerable overall military balance that would protect its more exposed allies even in conditions of non-nuclear inferiority, thus allowing them to resist Soviet intimidation—and also to dissuade the proliferation of national nuclear forces—the United States has had to extend the scope of its nuclear deterrence to cover its allies as well. If it were not for that, if the only requirement were the deterrence of nuclear attack upon the United States itself, then indeed all the labors of Rand, all the advocacy of additional strategic-nuclear strength, and a great part of the nuclear weapons themselves would always have been unnecessary.

By leaving the Soviet army out of the picture, and by constantly pretending that American nuclear weapons serve only to deter a most improbable Soviet nuclear attack upon the United States, it is easy enough to make the thinking of the nuclear strategist seem nothing more than self-serving, even mercenary over-elaboration. By leaving out the Soviet army and its armored division equipped, trained, and deployed to threaten invasion, one can ridicule any and every measure of nuclear force-building as useless and dangerous "overkill," a word which suggests an absurd excess measured by matching the many American nuclear warheads with the small number of Soviet cities—as if American nuclear warheads were in fact targeted upon Soviet cities to any important extent, and as if the United States could plausibly deter attacks upon other countries, and attacks not necessarily nuclear, by the desperate and unbelievable threat of destroying the cities of the Soviet Union. By leaving out the Soviet army, one can depict SAC and the rest of the military as mad bombers, forever wanting to burden the planet with more and more nuclear weapons, for no better reason than bureaucratic self-interest, or even for the profit of the contractors. Kaplan avoids such explicitly crude characterizations, and yet he implies them by his omission of any sustained and serious discussion of the non-nuclear imbalance of forces.

In reality, all the significant thinking of the Rand men and their many emulators, and almost all the building of new nuclear weapons, have been motivated by the very opposite of a quest for "overkill." Far from seeking to obtain more destructive power against the peoples of the Soviet Union, the aim rather has been to uncover ways and means of making retaliatory threats more selective, more flexible, and much less destructive—threats which could in turn plausibly deter Soviet threats themselves less catastrophic than the all-out nuclear attack upon America which is indeed most easily dissuaded by a very small nuclear force. As soon as it becomes necessary to do more, everything becomes far more difficult, and many more weapons far more accurate and resilient are needed.

Thus, for example, if the Minuteman missiles served only to deter a nuclear attack upon our own cities, then we should not be concerned by their vulnerability to attack. Even if the Soviet Union could destroy 90 percent of them with its own larger missile force, the remaining 10 percent would still be ample to destroy every Soviet city worth destroying—thereby still inhibiting any Soviet attack upon our cities.

But that is not the exclusive or even predominant purpose of the Minuteman force. It serves rather to add an element of deterrence in inadequate non-nuclear defenses. These land-based missiles, the most controllable of all long-range nuclear weapons, can best extend deterrence because they can be used selectively, perhaps in twos or threes, to threaten not Moscow or Kiev but military facilities relevant to the invasion to be deterred, facilities perhaps remote from any large civilian population.

Now the vulnerability of the Minuteman force does reduce its deterrent value. For how can the threat of using a *few* of the missiles be plausible when the Soviet Union can threaten just as plausibly to destroy all of them if any are used? Thus, ironically, to threaten less one needs more.

Those who flatly assert that the Soviet Union would promptly launch an all-out assault if it ever became the target of any nuclear attack at all, even one provoked by a massive invasion of Europe, are merely echoing a transparent Soviet attempt to inhibit the extension of deterrence. Of them we should ask whether the Soviet Union, having unleashed an invasion, would add the cataclysm too if, say, a single airfield were destroyed by a single warhead. And if not, what of two warheads and two airfields? And so on, in a reversal of "salami tactics."

Those, by contrast, who merely note that nuclear weapons are so destructive that any "selectivity" is dubious, and therefore exceedingly dangerous, have a far simpler and more impressive argument. But so long as they offer no better defense against the Soviet army than loose talk of new anti-tank weapons (the Syrians had thousands in the Beka'a against Israel in 1982), and still looser talk about the drunkenness of Soviet officers (drunk they fought, and drunk they won, in 1945 as in 1812), the only alternatives to the decline of Western security will continue to be nuclear forces resilient enough and powerful enough to extend deterrence, or else the enormous increases in defense spending that a truly adequate conventional deterrent would require.

When Kaplan reaches the "missile-gap" episode at the end of the Eisenhower era (yes, we meet yet again an avuncular Eisenhower, firm or almost so in resisting military demands), he becomes even more tendentious. We now encounter Air Force Intelligence enlivened by General George Keegan—the same Keegan who many years later introduced the entire rubric of beam weapons into our public debate, only to be branded as a fantasist by the Carter administration which was soon enough to fund those very weapons to the tune of hundreds of millions of dollars.

Kaplan recounts the fierce paper battles of the Air Force with the CIA experts who began to doubt, and later flatly denied, that there was any missile-gap at all. Kaplan has a good time in contrasting the frightening Air Force estimates of a hundred Soviet ICBM's by 1960 (and five hundred soon thereafter) with the tens forecast by the CIA, and the *four* actually known to have been deployed. But consider the circumstances: first, there was in 1957 the shock of the Soviet launch of Sputnik, the first man-made space vehicle; second, there was the plain fact that the booster which had launched Sputnik into orbit could more easily launch a nuclear warhead at the United States, thereby functioning as an ICBM—at a time when the United States had no such weapons; third, there was the carefully orchestrated Soviet propaganda campaign designed precisely to persuade the world that the Soviet Union was in fact producing ICBM's in quantity; fourth, there was the absence of reliable and comprehensive intelligence from inside the Soviet Union until the first satellite photography became available in 1960 (the earlier U-2 photography had been unable to bring back pictures from some of the most likely missile locations).

And then there was one more fact, the most significant of all, which Kaplan as it happens does mention, but only in passing and in another context, long after he has had his fun with Keegan and the Air Force. Although there was a controversy over the number of Soviet missiles of intercontinental range, there was *no* controversy about the medium-range ballistic missiles of which the Soviet Union had hundreds at the time. Thus the reader is invited on page 167 to ridicule Keegan's early estimate of hundreds of Soviet ICBM's, only to be casually informed on page 301 that the Soviet Union did indeed have hundreds of ballistic missiles, though not of intercontinental range.

To be sure, there was an absolute difference between the two classes of weapons so far as the vulnerability of American-based nuclear forces was concerned. But the ability to build medium-range missiles certainly suggested a parallel ability to build equivalent weapons of intercontinental range, and the Air Force was thus being only prudent in estimating that what could be build would be built.

Moreover, in the broader context of strategy, the difference between Soviet medium-range missiles aimed at Europe and ICBM's aimed at the United States was not all that great, given the fact that the purpose of American nuclear forces was (and is) primarily to protect Europe by deterrence. Obviously Soviet medium-range missiles that could destroy any European city diminished the assurance that American protection could confer.

The missile-gap was indeed a myth, if only because the Soviet Union was overambitious and tried to deploy too quickly a huge and most impractical first ICBM. But a myth has also been created about the myth—that the missile-gap overestimate was the result of unreasoned, groundless hysteria or even deliberate fabrication. In fact, it was induced by elementary prudence, for once misapplied. And there is worse, much worse: the missile-gap myth lasted only for three years, but the myth about the myth endures still. As do its evil consequences.

One such consequence is simply the inclination to disbelieve even the strongest evidence of increases in Soviet military power in any form. Another was confined to the inner world of the intelligence men themselves, but its effects were broad indeed, effects with which we must live, at great cost and some added risk: the systematic *underestimation* of the Soviet ICBM force from the mid-1960's onward. The same group of CIA analysts who exposed the overestimate of the 1957-60 period, and who thereby gained authority over the combined (or "national") estimates, systematically proceeded to underestimate the quality of new Soviet ICBM's and their rate of deployment, continuing to make the same gross error year after year until the mid-1970's. It is odd that Kaplan does not feel it necessary to censure error in the direction of imprudent optimism, though this was an error that persisted for many more years than the missile-gap error did, and which was altogether less excusable given the availability of satellite photography throughout the period.

It was on the basis of such profoundly misleading but ardently welcome estimates that Secretary of Defense Robert McNamara—traumatized by the Cuban missile crisis—cut off any major innovation in the American nuclear arsenal after 1963. Kaplan follows the fashion by celebrating, the post-1963 McNamara as the man who stood up to a voracious military and especially the Air Force (which wanted to build not only a new bomber but also the WS-120—just the sort of large ICBM that the Soviet Union now has in its SS-18). Kaplan's account eventually extends to the McNamara of delicate personal agonies about Vietnam who was to expiate his sins by serving as the unlikely preacher for the poor at the World Bank. But there is one aspect of McNamara's doings that Kaplan does not choose to discuss, namely, that his openly declared policy of one-sided restraint was based on a gross miscalculation of the Soviet response—an error of epic proportions, unambiguously documented in the successive Annual Defense Reports for Fiscal Year 1964, 1965, 1966, and 1967.

In fact, the systematic underestimation of Soviet armaments went far beyond the ICBM's alone and lasted long beyond McNamara's tenure at the Pentagon, serving as the foundation for policies of arms control and non-deployment which might have worked well if only our antagonist were Great Britain or perhaps Sweden. It was those policies which engendered the predicament of today, in which a weapon as unsatisfactory as the MX is nevertheless a greatly belated necessity, and in which nuclear weapons must awkwardly be deployed in the crowded setting of Europe to serve as costly, inadequate, but essential substitutes for the intercontinental nuclear advantage that was so frivolously surrendered.

Kaplan concludes his book by proclaiming that "the bomb" remains "a device of sheer mayhem, a weapon of cataclysmic powers no one really had the faintest idea how to control. The nuclear strategists had come to impose order—but in the end, chaos still prevailed." If this were said in order to call for more thought and better planning, for further refinements of engineering and for new and better policies, then the matter might be debated on its merits. But that is not at all the case. Kaplan relishes the chaos he has placed in prospect, and insists on the futility of any remedy.

And yet it is a fact most easily documented—without recourse to the Freedom of Information Act—that today's nuclear weapons are far less destructive and far more reliably controlled than those of twenty years ago; and it is a fact that needs no documentation at all that we have lived since 1945 without another world war precisely because rational minds did not surrender to unreason but rather devised the plans and procedures, the deployment modes and the policies that extracted a durable peace from the very terror of nuclear weapons.

Review of *Years of Upheaval*, by Henry Kissinger

Metternich, Bismarck, Disraeli, the classic trio of great "foreign ministers," were not in fact foreign ministers but chief ministers, heads of their respective governments; and Talleyrand too was in practice a chief executive at the time of his main achievement, at the Congress of Vienna. For some foreign ministers properly defined, the example of those figures has been unfortunate, tempting them to try to steer the ship of state without having its command. Unable to include domestic affairs within his own sphere of responsibility, lacking in authority over the military chiefs, the foreign minister who is no more than that must depend on the chief executive to sustain his policies—often while being his rival, real or suspected, and correspondingly likely to be undermined at crucial moments.

No wonder then that those foreign ministers who have tried to forge policies as deep and as broad as those of the classic trio, or have sought merely to imitate their fancy diplomatic footwork, have been doomed to failure. Schemes of policy aborted halfway, or just as disastrously diverted into unintended paths, and a diplomatic tone made up of dissonant voices fatally off-key (obstinate when firmness was the aim, or merely weak when a yielding resilience was intended)—those have been the usual results of ambitious ministerial foreign policies. Wiser men placed in that office, or those simply less ambitious have been content to manage affairs from day to day, with no real scheme of sustained action. Such ministers are held in high esteem by their departmental officials, since foreign ministries everywhere are structured not to make policy but rather to avoid any departure from continuity.

In such cases, the foreign minister is merely the chief administrator of his country's diplomacy, the executor of policies made by others (who may know too little of world affairs to set the right goals); or else he becomes the keeper of established policies, shaped by the circumstances of the past and lovingly preserved. That may seem a prudent enough course to follow, and so it is—but only for the minister himself, who can thus easily avoid blame for the failure of any new initiative. For his country and government on the other hand, a policy that preserves continuity because of sheer inertia can easily turn out to be very costly, and may even become highly dangerous, since powerful forces far more active than foreign ministries are loose on the world scene, and these will demand either some adjustment, or else energetic reactions.

With modern states so organized that foreign affairs are the concern of specialized bureaucracies, and with the enormous growth in the domestic activities of all governments, the stage is set for the systemic failures of foreign policy characteristic of our century of wars: the press of domestic matters prevents the chief executive from devoting sufficient attention to foreign affairs while, on the other hand, the foreign minister lacks the wide-ranging authority that the exercise of his function properly requires, notably control over military policy. This systemic defect has been a major factor in the spectacular foreign-policy disasters of our times: August 1914, shared by all and notoriously precipitated by the crippled departmental diplomacy pursued by Britain, France and Germany; the inter-war French failure that resulted from the fatal disharmony between an alliance-building foreign policy based on the lesser powers around Germany, and a military policy which renounced the offensive capability needed to protect such lesser allies (which could jointly have diluted German strength on many fronts in May 1940, if the French had protected each in its own moment of danger); and since 1945, with war betwen the Great Powers duly avoided by the awesome fears that nuclear weapons so beneficially evoke, Great-Power wars against lesser enemies—notably the Anglo-French Suez adventure of 1956 and, on a far greater scale, America's war in Indochina.

In the case of Suez, the Foreign Office and the Quai d'Orsay were left as the impotent spectators of (British) military planning that ignored the realities of world politics: if the deed could be done at all, it could only be done swiftly, and not by the leisurely process of a full-scale amphibious landing mounted by an armada that steamed slowly across the full width of the Mediterranean (Eden's premise was the easy removal of Nasser; the Military premise was that the Egyptians could only be defeated by large-scale war—and the blatant contradiction was not allowed to disturb the decision). As for Indochina, there the worthy aim of resisting Hanoi's imperialism was perverted by all the follies of excess and bureaucratic self-indulgence that a luxuriously well-supplied military structure could devise, because there was only McNamara's scientistic misunderstanding of war to provide it with guidance, instead of a coherent foreign policy from which purposeful military directives could be obtained. When such a policy was finally achieved under Henry Kissinger, Hanoi was brought to the very edge of capitulation, being saved only by the unreasoning domestic opposition to the war which all the errors of the past had by then engendered.

We can therefore recognize the most important factor that enabled Kissinger to become a true successor to the great "foreign ministers" of the past: thanks to the combination of Nixon's virtues and of his great weakness, Watergate, Kissinger was able to act more or less as a chief executive during 1973-74; his elevation to Secretary of State on September 22, 1973, merely registered—and grossly understated—his effective control over America's external conduct. To be sure, with a president sinking into impotence as a result of the daily unfolding of the Watergate scandal, any respectable and competent figure could have enjoyed security of tenure as Secretary of State, since his resignation would have inflicted another great wound. But if Kissinger had been merely that and no more, the chances are that he would have achieved little or nothing in his office, since the potential for departmental anarchy built into the American system would then have asserted itself; without a lively presidential effort, each department will naturally tend to stifle the initiatives of all the others. The system of "checks and balances" of which the Americans are so proud was meant to apply to the branches of the government and not as between the departments of the executive branch, but it is manifest in great strength within its confines; thus the Americans can boast of the world's most elaborate machinery for immobilism in foreign affairs, especially now that each huge bureaucracy can exploit for its own obstructionist purposes that whole jungle of restrictive legislation that Congress originally created to make itself more powerful in foreign affairs.

But Kissinger, by reason of his personal authority and bureaucratic cunning, received much more than his share of the power that ineluctably flowed out of Nixon's afflicted White House, becoming steadily more powerful within the executive branch just as the executive *in toto* was losing ground to Congress. By the end, of course, Kissinger's preeminence had become hollow, since the executive he was able to dominate was itself becoming impotent. But in the interval, before the second devolution caught up with the first, Kissinger had the opportunity to act more freely than any other modern Secretary of State. *Years of Upheaval,* the second volume of his memoirs, tells us what he made of the opportunity. And he did so much that the 1,214 pages of the book (not counting notes, index and a documentary appendix) are scarcely excessive, even though they cover a period of only nineteen months or so.

The first substantive chapter of *Years of Upheaval* reviews the deteriorating state of Indochina, afflicted by the relentless pressure of the North Vietnamese and their then allies on the ground, and by the equally destructive consequences of American domestic opposition to the war, by then manifest in the form of congressional budget-making and restrictive legislation. Less than a decade had passed since those days and yet it is already very difficult to credit one's own memory: did so many academics, journalists and politicians really believe that the Viet Cong was an autonomous entity, dedicated to national liberation? Did they truly regard the Khmer Rouge as an improvement on Lon Nol's regime? Did they actually consider Hanoi's rulers to be men of benevolent temper? Of course they did—and since the deluded and the deceivers are still very active on the American scene, the mystification must be perpetuated in one form or another, to protect reputations made or amplified by opposition to the war.

In a society that is forgetful as well as forgiving, names that ought to evoke scorn still claim respect: one thinks of Richard Falk among the academics (not to speak of the deservedly forgotten band of the "Concerned Asia Scholars"), of Harrison Salisbury and the author of *Fire in the Lake* among the publicists; and then, of course, there is that whole crowd of columnists and reporters whose professional standing was acquired in the days when Nguyen Van Thieu was equated with

Hitler and Le Duc Tho was presented as a latter-day Jefferson. This was the guilty elite that greeted William Shawcross's *Kissinger, Nixon and the Destruction of Cambodia* with such purposeful enthusiasm, since its distortions and documentary manipulations (relayed in adulatory columns and lengthy, uncritical reviews) served so well to obfuscate the obvious, namely that opposition to the war in Cambodia resulted in the victory of a régime which was, quite simply, homicidal. Conclusive evidence that the leaders of the Khmer Rouge acted by long-standing design was disregarded by Shawcross, who continually insinuated that it was the American bombing (of areas largely uninhabited, but used by North Vietnamese troops) that somehow transformed "agrarian reformers" into assassins. Only people with a very guilty conscience and with a whole past to live down would have fallen for such myth-making, but then of course those characteristics do define quite accurately a large slice of today's opinion-making elite in the United States: hence Kissinger's documentary appendix, which is meant to expose some of the more glaring distortions in the Shawcross book.

The chapter itself is a record of Kissinger's visit to Indochina. Much of it deals with his time in Hanoi, where he encountered leaders for whom the Paris Agreement was merely a stepping-stone to further war, and war to the finish. Readers of the first volume of these memoirs will already have made the discovery that Kissinger can write rather well, having made the brief character sketch something of a speciality: If Kissinger has a weakness as a writer (a weakness which may perhaps also be reflected, if only dimly, in the doings of the practitioner) it is that he tends to magnify the character of his subjects, and especially of his antagonists, so that Pham Van Dong the provincial-minded bureaucrat of violence becomes the dedicated revolutionary, and fanatical stubbornness becomes an implacable tenacity of historic proportions. Admittedly all these things are perfectly congruent, and yet anyone who lives out his life as Pham Van Dong must be mean-spirited above all, and that is not the impression given by the text.

As for the substance of the matter, Kissinger obviously tried to induce the North Vietnamese to see the advantages of an accommodation for the economic reconstruction of Vietnam and the welfare of the Vietnamese people, but of course he failed. As Lee Kuan Yew is reported to have said when the North Vietnamese faced the choice they went for the soft option—and continued the war.

The next chapter is also set in Asia and also centered on a journey, this time to Peking, where Kissinger had meetings with Zhou Enlai (the author had adopted Pinyin and Mao Zedong. As usual, Zhou was expansive and Mao enigmatic, and as usual Kissinger was beguiled by Zhou's discourse and excited by Mao's allusive chit-chat. Some business was also transacted, namely the agreement to set up liaison offices in Washington and Peking, but it is clear that the visit was an anticlimax. By now, most of what could be achieved had already been achieved. The opening to Peking had already served the Nixon administration very well, by feeding dramatic imagery to the public and especially by soothing left-wing opinion which in those days still loved Mao; it was also useful diplomatically (up to a point) in dealings with the Soviet Union. And the Chinese could also be serviceable in saying helpful things to such countries and political groups as were willing to listen to them—several lesser African countries and the Japanese Socialists for example.

But there was never any substance to the widespread expectation that Sino-American cooperation could be made to yield great results in the realm of strategy and security. China's military weakness, and in particular its lack of offensive capability, deprives the "China card" of any real strategic value. To help NATO in a Soviet war the Chinese would have to *attack* the Soviet Union and the Chinese have no such capacity; more to the point, China cannot help to deter the Soviet Union on any front because the latter now has enough disposable military power to keep both China and NATO on the defensive, while using its vast *masse de manoeuvre* to threaten either, or else to make war on a third front. Having made the vast investment required to build bases and infrastructures all along the Chinese border, having greatly increased its forces in all categories, the Soviet Union has in fact absorbed the full strategic consequences of China's hostility. The cost thus imposed has brought advantage to the West, but no amount of friendly diplomacy between Washington and Peking can now increase that benefit—nor for that matter could the benefit be lost if the diplomatic climate were to deteriorate.

Kissinger could have easily negotiated full diplomatic relations with Peking on the terms that the Carter administration was to accept. That he resisted that easy diplomatic success stands very much to his credit, and that of Nixon too: the diminished caliber of Kissinger's successors was perfectly demonstrated by their eager self-congratulations over the so-called normalization—achieved by the simple expedient of accepting Peking's terms.

The chapter on "The Year of Europe" received its title, one presumes, in a spirit of irony, for of course there was no year of Europe. Having spent much time in intimate dialogue with adversaries, Kissinger was being accused at home and abroad of withholding his attentions from the Allies. He responded by planning a suitably dramatic initiative, which was to be introduced by a self-designated "major speech" about inter-Allied relations, and was to culminate in the solemn signature of new charter or treaty between the United States, the NATO Allies and the Japanese. In other words, it was to be "The Year of the Allies." But the Allies for all sorts of reasons would not cooperate, and some had outright sabotage on their minds: the British with Heath at the helm wanted to toe the French line at all costs, or at least to be seen to do so; the Italians and Japanese were evasive; the Germans under Brandt were determined to show how successfully they could resist American diplomacy, even if there was nothing contrary to their interests to resist; and the French could not in the end be other than equivocal because they were represented by Michel Jobert, who sought to become Pompidou's successor by posing as the neo-Gaullist defender of an independence that the Americans had by then finally learned to appreciate, and which they had no intention whatever of compromising—least of all by a new Atlantic Charter that would largely have reaffirmed *American* obligations.

Soviet-American relations are, implicitly at least, the pervasive theme of the whole book, but there are no striking new insights nor any interesting revelations when Kissinger deals with the subject directly. Soviet matters become interesting only in other contexts, whether the Chinese or the Middle Eastern. The Soviet leaders could not of course understand Watergate. Even much later, no amount of lecturing by their "consultants" and Americanologists could persuade them that the American president could not, for example, force passage of SALT II through a reluctant Senate, so that one may imagine their utter inability to understand how Nixon could be seriously embarrassed by the Watergate investigation. But in due course they did at least absorb the hard fact that Nixon could not, for example, deliver on the implied promises of vast capital loans, and their reaction was to withdraw into a more reserved attitude. Kissinger records this reaction but by its nature there is not much to write about.

The Egyptian and Syrian attack on Israel of October 6, 1973, started a new crisis whose consequences soon proved to be exceptionally wide-ranging. Kissinger, seemingly from the first, saw the crisis as a great opportunity and his account of what followed takes up virtually the whole of the second half of the book. The encounters with Sadat, the Israelis and President Assad of Syria are described in great detail and at great length but the material is fascinating and fully warrants Kissinger's extended treatment. The purpose of his shuttle diplomacy was, of course, to bypass each side's foreign-affairs bureaucracy and deal directly with the principals, who had a greater freedom of action and a greater readiness to explore imaginative solutions than their underlings. The world is full of Middle-Eastern experts and "Arabist" diplomats who insist that Kissinger did not tackle the "real issue," namely the Palestinians. The assertion that the Arabs would never accept such an accommodation with Israel used to be the stock-in-trade of such experts; when some Arabs, at least, let them down by behaving as realistic national leaders prepared to acknowledge the facts of power, such experts took refuge with the Palestinians, whose intransigence was far more reliable. For it was not just Egypt that took the road of accommodation in 1973, but Syria also. In the wake of Kissinger's diplomatic bridge-building, "rules of the game" were established for the first time between Syrians and Israelis, whereby each side in its dealings with the other has since been using force in a limited and controlled fashion. It seems that the minimum of reciprocal understanding that Kissinger engendered between the two sides in 1974 has never quite evaporated.

Kissinger's Middle East policy—operated in a far wider arena than that of the Levant, if only because the Soviet Union kept trying to re-enter the ring (without, however, doing the one thing that would have earned its admission, namely re-establishing relations with Israel), while Kissinger himself kept trying to enlist other Arab powers—notably the Saudis—in support of his efforts. And then of course the Allies kept pressing for quick results—without, however, being willing to contribute anything to achieve a settlement. The great failure was of course the passive acceptance of the oil-price revolution, whose disastrous economic consequences have continued to be felt ever since. Kissinger has been accused of having actually encouraged the original, small, pre-1973 increases extorted by the Shah of Iran but the most that can be charged against him after 1973 is that he failed to appreciate how crippling the consequences of OPEC's price increases would be for the productive economies of the world.

So far it has been Kissinger's destiny that his achievement has only been magnified by the doings of his successors. Carter's foreign policy, the feeble probity of Cyrus Vance and the goings-on of Kissinger's parodic successor as National Security Advisor, could only enhance Kissinger's residual personal authority on the world scene—and his reputation has been further consolidated by the Reagan administration's record to date. Carter's men were fiercely determined to do the opposite of whatever Kissinger had done or would do; Reagan's first team—increasingly dominated by Haig, whose power over policy kept increasing to the very day of his fall—embraced rather than avoided imitation, and made a very bad job of it (the Buenos Aires-London shuttle was only the most vividly absurd example). Where once there was Kissinger on the scene moving in a sinuous diplomatic dance, we saw instead the clumsy tread of the elephant.

Kissinger's book may be read as a political travelogue; as a major contribution—although naturally highly subjective—to contemporary history; as a first-class manual of statecraft; and a practical guide-book to jet-age diplomacy. It is a thorougly good read.

Review of *Power and Principle: Memoirs of the National Security Adviser, 1977-1981,* by Zbigniew Brzezinski

Zbigniew Brzezinski's book is an honest and well-written account which will be valuable to historians and attractive to many readers. Inevitably, it is also a record of failure—not his own personal failure, of course, and not necessarily the failure of the nation's foreign policy either (for it is only in the retrospect of decades that the ultimate results of the doings and undoings of a world power can truly be known) but certainly the political failure of the Carter administration, in whose 1980 defeat at the polls foreign affairs played a most unusually large part.

The National Security Adviser's major role is to protect the president's interest in the conduct of external affairs. What may seem desirable to the State Department, the Arms Control Agency, the Pentagon, or to an "inter-agency" committee of all three may still be damaging to the president, and it is the National Security Adviser who is supposed to protect him in such circumstances; in addition, the National Security Adviser can also attempt to protect the president from himself—as Henry Kissinger apparently did with some frequency especially in the last phases of Richard Nixon's presidency.

Brzezinski as National Security Adviser obviously failed to protect Jimmy Carter. Without manifesting a resentful disloyalty to the man who placed him so high, without perhaps being fully conscious of doing so, Brzezinski explains throughout the book why Carter could not be protected from the departments, notably State, and why above all he could not be protected from himself.

A classic example of a very clever but basically unintelligent man, Carter was incapable of understanding the fundamentals of international affairs even while acquiring much detailed knowledge of their mechanics and surface manifestations; and since he never realized how much he was in need of help, he would not let Brzezinski guide him, either. Nixon, when driven by violent emotions, would sometimes want to do foolish things; but if his orders were simply ignored for a brief interval by Kissinger, no harm would follow because his rational judgment would reassert itself as soon as he calmed down. Carter was much better at controlling his emotions but unwisdom was his permanent condition, revealed in foreign affairs as soon as he chose Cyrus R. Vance as his Secretary of State and finally exposed in full view by the Iranian crisis.

One example will suffice. On November 10, 1979, there was a meeting (without the President) at the National Security Council to discuss the possibility of expelling the Shah from the United States. Vance and Vice-President Walter Mondale favored explusion, while Defense Secretary Harold Brown suggested that the Shah be prevailed upon to announce his intention to leave once he recovered from his illness. Brzezinski opposed this concession to the "students" who had seized the U.S. embassy and its staff in Teheran the week before. He quotes himself at the meeting: "A month ago we backed down to the Soviets and the Cubans after declaring that we found the status quo [the presence of a Soviet brigade in Cuba] unacceptable. Now we shall back down again. What will this mean for our international role as a global power? Who will find us credible hereafter?" When the question was referred to the President, he "flatly" pronounced against expulsion.

In making this decision Carter had apparently showed that he understood the foreign-policy content of the issue—that is to say, the worldwide loss of American authority which an expulsion would entail. But not at all. In a matter of days (on November 14) Carter reversed himself, and decided that Mexico should once more be asked to receive the Shah. The Mexicans, reflecting in their conduct the very loss of American authority which Carter's decision brought about, naturally refused to accept the Shah (who was eventually consigned to the extortionist Omar Torrijos of Panama).

Brzezinski does not report why Carter had earlier "flatly" rejected expulsion. It cannot have been because he truly understood what was at stake, for then he could not possibly have changed his mind so soon (no new relevant facts had intervened). Carter's reasons must have been of a lesser order, perhaps personal pique at being bullied by the Teheran "students" or misplaced optimism concerning the early release of the hostages.

In any event, in a pattern that was repeated incessantly in his administration, Carter made himself an expert on all the details of the hostage question, while failing to grasp the fundamentals, namely, that a world power must always consider the worldwide effects of its action in any particular setting. In the circumstances, this meant that the Teheran anti-Americans had to be confronted, not appeased.

In one case, to be sure, the peculiar defects of Carter and his foreign-policy team had a most productive outcome: had it not been for the evident intention of the Carter administration in 1977 to reconvene the Geneva conference on the Middle East under joint American and Soviet sponsorship, President Sadat would never have decided to go to Jerusalem. The Egyptian president had risked his neck to get the Soviet Union out of Egypt; he had built his entire policy on a deliberate rejection of the Soviet alliance and on a reassertion of Egypt's national identity (as opposed to Nasser's pan-Arab fantasies). When Sadat discovered that the Americans, of all people, wanted to cast Egypt back into the Arab fold, and under the patronage of the Soviet Union as co-chairman of the Geneva conference, he had the courage and inspiration to break out by creating his own context of negotiation, directly with Israel. The risks of going to Jerusalem were great, but the spectacular unwisdom of American policy entailed even greater dangers: at Geneva, Egypt could only expect to find itself outflanked by the Syrians and blackmailed by the Russians.

Then came the prolonged bilateral and trilateral negotiations which eventually resulted in the peace treaty between Israel and Egypt. Cyrus Vance's contribution to these negotiations, which Brzezinski generously stresses—his treatment of Vance spells out all their differences, but in a most gentlemanly fashion—owed a great deal precisely to the quality that made Vance unfit for his office. The kindly man who in dealings with Iran would not sanction the use of force to protect American interests, was kind to Egyptians and Israelis alike, infinitely attentive to all the nuances of their interests. As always he concentrated relentlessly on the single issue at hand, always searching for every possible avenue of accommodation, treating Egypt and Israel as if they were two well-paying clients of his law firm, and never as client-states of the United States of America.

As for Carter, he most certainly deserved a Nobel Prize for attending so assiduously to the negotiations which he had so strongly promoted. Both the Egyptians and the Israelis were astonished to discover that Carter was willing, over a very extended period, to immerse himself in the most minute negotiating details. It would be churlish to deny Carter's deep-seated commitment to peace, but it is also plain that his attitude to the negotiations owed much to the fact that once again he never understood the fundamentals. Carter saw himself throughout in the role of a mediator. Yet it was not his diplomatic abilities that the Egyptians and Israelis wanted but rather his commitment of American prestige and resources—which would greatly add to each side's "take" from the negotiations. Sadat and the Israelis could have made peace on their own, but they would have been foolish to do so: each impasse in the negotiation was resolved as much by American contributions as by reciprocal concessions.

Brzezinski writes a great deal in this book about the "policy process." Again we are reminded of the extent to which the elaborate machinery of policymaking, which is supposed to coordinate the large and undisciplined departments of the executive, produces paralysis or confusion as its most natural result. The Constitution calls for a separation of powers between the branches of government; it cannot be blamed for the "checks and balances" which have arisen *within* the executive as a consequence of luxuriant bureaucratic growth. Brzezinski notes, for example, that the normal consequence of overruling the State Department was a hostile leak, a very effective, if certainly improper, "check." Kissinger simply accepted that the "process" did not work, and that it had to be circumvented if serious initiatives were to be launched. Brzezinski could not or would not emulate his predecessor. Hence his impact on the making of policy was small.

One instance of confusion is worthy of special mention. In January 1979, with the Shah still in Teheran and the Bakhtiar government about to be formed, General Huyser was to be sent to Iran to make direct contact with the military chiefs. A meeting was convened on January 3 to decide what instructions Huyser should be given. Vance, Deputy Secretary of State Warren Christopher, and Mondale insisted that the military chiefs should be warned against a coup. Brzezinski agreed, but argued that Huyser should also encourage them to stage a coup "in the likely event that Bakhtiar should fail." In practice, therefore, Huyser received directly contradictory instructions. The predictable effect was to paralyze the Iranian military, the one group that might yet have saved the situation.

The next day, Brzezinski met with the President in Guadeloupe, where a four-power summit was being held. The scene was characteristic: Carter in a bathing suit in his cottage, sitting on an icebox, Hamilton Jordan also in a bathing suit, sprawled on a sofa. (Such scenes, which occur frequently in the book, can easily persuade one that formal attire has its uses after all.) Vance was on the phone, in "considerable agitation" because the Iranian military had told U.S. Ambassador William Sullivan that they wanted to keep the Shah in Teheran and suppress the rebellion in full force, with as much bloodshed as it would take. Vance and Mondale wanted permission to tell the military chiefs that the United States would oppose their use of force. Brzezinski joined in the argument, which "lasted a long time." Finally, he records: ". . . much to my satisfaction . . . the President took a very firm line. He told Cy that he did not wish to change General Huyser's instructions."

This "decision," however, merely reaffirmed the earlier contradictory instruction—Huyser's yes/no to a coup. Obviously, the Iranian military chiefs had not approached Sullivan because they were in need of conversation. With mobs on the rampage and the Soviet Union in malevolent proximity, they were seeking a firm and unequivocal U.S. endorsement of a dangerous but potentially decisive act. To add the impact of military choreography to mere words, U.S. aircraft carriers could have been sent into the Gulf, military supplies (needed or not) could have been airlifted. Instead, there came from Washington only conditional statements, full of ifs and buts.

Even in retrospect, Brzezinski does not seem to recognize what had to be done to implement the policy which he so eloquently promoted. For after the Guadeloupe episode he notes, with apparent surprise, "Alas, nothing happened. Vance conveyed the instructions orally to Sullivan, and I have no doubt that he did so faithfully. At the other end, the military . . . simply procrastinated."

Toward the end of this book, Brzezinski produces a formal list of what he loyally defines as President Carter's accomplishments in the areas of foreign policy and defense. The list encapsulates the entire problem of the administration, and of this book. Naturally, it includes the Camp David peace agreements and the Panama Canal treaty, along with the post-Afghanistan "Carter Doctrine," and so on. But the first item on the list is the Carter human-rights policy, which as it happens left the Soviet Union more repressive than ever while doing in a number of friendly rulers, and the last item is the SALT II agreement, which remained unratified. In many ways, and not only in its honesty, this is an innocent book.

14

Delusions of Soviet Weakness

In recent years, entire books have appeared which argue that the Soviet armed forces are much weaker than they seem. Citing refugee accounts or personal experience, they depict the pervasive technical incompetence, drunkenness, corruption, and bleak apathy of officers and men. Drunken officers and faked inspections, Turkic conscripts who cannot understand orders in Russian drowning in botched river-crossing tests, the harsh lives of ill-fed, ill-housed, and virtually unpaid Soviet conscripts, and a pervasive lack of adequate training fill these accounts.

It is odd how all these stories (each true, no doubt) contrast with the daily evidence of the routine operations of the Soviet armed forces. Merely keeping its warships seaworthy and supplied in distant and often stormy waters demands a great deal of discipline and expertise from the officers and men of the Soviet navy. Even more skill is needed to carry out successfully the missile launches and gunnery trials that are also part of the Soviet naval routine. Likewise, we have the evidence of Soviet air operations; they too require a great deal of competence, both in the daily training sorties of the fighters and in the long-range flights of the bombers and transports.

Nor can the Soviet army fake all the disciplined maintenance, tight planning, and skills needed to assemble, move, and operate the many thousands of complicated armored vehicles, hundreds of helicopters, and countless smaller weapons in its exercises. It only takes a little drunken inattention or technical incompetence, or mere apathy by maintenance crews, to cause an aircraft to crash; a little more can sink a ship; and the delicate gear box of a battle tank is easily wrecked.

It is true that at fairly regular intervals we learn of spectacular failures in the upkeep of the Soviet armed forces. Breakdowns at sea lead to much photography of submarines adrift in the ocean and to much speculation over possible radiation leaks. Word of plane crashes reaches us now and then, and most recently there was solid evidence of

229

huge explosions in the weapon stores of the Northern Fleet in the Kola peninsula. It is perfectly probable that Soviet standards of maintenance are lower than those of the United States, but the difference is scarcely of dramatic consequence. All armed forces, including those of the United States, have their collisions, their air crashes, their catastrophic breakdowns. The Soviet armed forces may well have more than their share. Yet it was never by superior efficiency that first Russia and then the Soviet Union became so very powerful, but rather by a combination of numbers, persistent strategies, and a modest technical adequacy.

When the actual record of war is assessed, not from official accounts but from the testimony of those who were there, it becomes quite clear that battles are not won by perfection but rather by the supremacy of forces that are 5-percent effective over forces that are 2-percent effective. In peacetime, when all the frictions of war are absent, when there is no enemy ready to thwart every enterprise, effectiveness may rise to dizzy levels of 50 or 60 percent—which means, of course, that filling in the wrong form, posting to the wrong place, supplying the wrong replacement parts, assigning the wrong training times, selecting the wrong officers, and other kinds of errors are merely normal. Matters cannot be otherwise, because military organizations are much larger than the manageable groupings of civilian life that set our standards of competence; and because their many intricate tasks must be performed not by life-career specialists like those who run factories, hospitals, symphony orchestras, and even government offices, but by transients who are briefly trained—short-service conscripts in the case of the Soviet armed forces.

Actual alcoholism, in the severe, clinical sense, is now epidemic in the Soviet Union, where so many lead bleak lives, no longer alleviated by the once vibrant hope of a fast-approaching better future. So drunkenness is no doubt pervasive in the Soviet armed forces. But Soviets have always been great drinkers. Drunk they defeated Napoleon, and drunk again they defeated Hitler's armies and advanced all the way to Berlin. All these stories of corruption are also undoubtedly authentic. But no great military empire is likely to be undone by generals who procure villas through corrupt dealings, nor by sergeants who take the odd ruble off a conscript; Anglo-Saxon morality makes much of these things, history much less. Corruption in the higher ranks can demoralize the troops—but not if it is accepted as a normal part of life.

On the question of loyalty, even less need be said. Should the Soviet Union start a war, only to experience a series of swift defeats, it is perfectly possible that mutinies would follow against the Kremlin's oppressive and most unjust rule. But if the initial war operations were successful, it would be foolish to expect that private disloyalty would emerge to undo victory and disintegrate the armed forces. There will always be a small minority of lonely heroes with the inner resources to act against the entire power of the world's largest and most complete dictatorship. The rest of us weaker souls will stay in the safety of the crowd—and the crowd will not rebel against a uniquely pervasive police system at the very time when successful war is adding to its prestige, and the laws of war are making its sanctions more terrible.

Only one claim can be allowed: it is true that the ethnic composition of the Soviet population is changing, with non-Russians making up an increasing proportion of the total. This creates problems of loyalty that are unknown in the United States, because in the Soviet Union, distinct nationalities persist with their own languages, ethnic senti- ments, and sometimes strong antagonism to the Russian master- people. As the proportion of non-Russian conscripts increases, lan- guage problems also increase, and because many of these conscripts come from backward nationalities, they are harder to train in modern military skills, even if they do know the Russian language. There is also a greater potential for ethnic strife, already manifest in barrack- room fights.

In the *very* long run it is possible and even likely that the non- Russians, or at least the larger non-European peoples—the Uzbeks, Kazakhs, Tadzhiks, and so on—will demand full national indepen- dence and struggle for it, eventually causing the dissolution of the Soviet empire, which is the last survivor of the European empires that dominated much of the entire world until a generation ago. Demogra- phy is indeed a powerful and relentless force, but slow in effect. In 1970, out of a total Soviet population of 242 million, 74 percent was Slavic and 53 percent actually Russian. (Some of the fiercest antago- nisms are between Russians and other Slavs.) In that year, there were 35 million people of Muslim origin (mostly Turkic), just under 15 percent of the Soviet total. By the year 2000 it is projected that the Muslims will account for more than a fifth (21-25 percent) of the total population of 300 million, with Russians at 47 percent and all Slavs at 65 percent. Naturally the change will be felt sooner and more strongly in the younger groups of military age. For example, out of the 2.1 million males projected to be at the conscription age of eighteen in 1985, the non-Slavs will account for more than 35 percent, and quite a few of them will not know enough Russian to obtain the full benefit of training.

But of course the armed forces of a multi-national empire know a thing or two about managing diverse nationalities. Those with a high percentage of dissidents, such as the Estonians, Western Ukrainians, and Jews, can be safely employed in military-construction battalions, which are virtually unarmed, or in other support units far from combat; those with many illiterates or conscripts whose Russian is poor, such as the Kirghiz, Turkmen, and Tadzhiks, can be placed in the unde-manding mechanized infantry of second-line divisions. There are problems, but they remain quite manageable. The real problem of national self-assertion is for the distant future.

So far nothing precise has been said about the most obvious attribute of the Soviet armed forces: their sheer numerical strength. The gross totals are well known, and mean little. As against the 30 large divisions of the U.S. army and marine corps, active and reserve, the Soviet army has 194 divisions, smaller by a third on average but just as heavily armed. One-third are fully manned, one-third are half and half, and the rest are mostly manned by reservists—but all Soviet divisions are fully equipped, even if not with the latest and best, and all have a full-time professional cadre, even when their line units are manned by reservists. The Soviet tactical air force has some 6,000 strike aircraft, fighters, and fighter-bombers, less advanced on average but also of more recent vintage than the 5,600 or so equivalent aircraft of the U.S. air force, navy, and marines. Another 1,250 interceptor-fighters serve in the territorial air defenses (along with more than 9,600 anti-aircraft missiles), and the Soviet navy's land-based aviation also includes some fighter-class aircraft.

For the Soviet navy, one ship list prepared by the U.S. navy shows 1,324 "surface combatants," as against its own 285 surface warships; 367 submarines, as against 99; and 770 auxiliaries, as against its own 105 logistic and support ships. The figures are of course grossly inflated, but even the most sober count that excludes the old, the inactive, and the small would still list 290 major Soviet surface war-ships, 119 nuclear and 157 diesel-attack submarines, and 360 land-based naval bombers, of which 100 are modern machines of trans-oceanic range.

No true military balance is made of mere lists, however. The place and the time, the allies present on each side, and the circumstances of the nation and of the particular theater of war will govern what can be achieved, and indeed what forces can be deployed at all. No estimate can be made for expeditionary ventures in undefined theaters of hypothetical war—except to say that the power of the Soviet military wanes drastically as the distance from the Soviet Union increases, much more so than does that of the American forces, which are far better equipped to reach and fight in faraway places. But we can make rather solid estimates for the continental theaters of war directly adjacent to the Soviet Union—in Europe, the Middle East, and northeast Asia. The results are grim.

In the five possible war theaters of the North Atlantic alliance—northern Norway, the "central" front in Germany, northeast Italy, the Thrace frontier of Greece and Turkey with Bulgaria, and the Turkish border with the Soviet Union in remote eastern Anatolia—it is clear that the ground forces of both the United States and its allies, those already deployed in peacetime and those to be mobilized, would be outnumbered, outgunned, or both. By adding absolutely everything on the books—including Turkish infantry and the American National Guard, in addition to the manned forces actually in place—the total number of alliance divisions for the five theaters rises to 144, as against a combined Warsaw Pact total of 170. That is scarcely a catastrophic imbalance, and the situation looks even easier for the alliance when we recall that the Warsaw Pact total includes the divisions of rebellious Poland, unwilling Hungary, restive Czechoslovakia, doubtful East Germany, and uncooperative Rumania.

If we make a somewhat finer comparison, however, including only tank and mechanized divisions on the alliance side, thus removing a mass of ill-armed and immobile infantry forces of low military value, while at the same time eliminating *all* the non-Soviet forces of the Warsaw Pact, 80 divisions of the alliance remain, while the Soviet army alone has 109—*after* leaving 78 Soviet divisions to face the Chinese border, occupy Afghanistan, and control Iran's long border. These 109 Soviet divisions are smaller than the Western divisions, but no longer by much, and they do belong to one army under one central authority, whereas the Western total is split among the armies of the United States, Canada, Britain, Norway, Denmark, West Germany, Holland, France, Portugal, Italy, Greece, and Turkey—and the French divisions are not under alliance command and not necessarily available, the Greek divisions are of uncertain allegiance, and the American reserve forces must first be mobilized, then filled out and updated in training, then transported across the ocean.

If we include the nonmechanized forces of high military value (such as the American and Soviet airborne divisions), and exclude alliance forces not rapidly available for reinforcement, the realistic alliance count is on the order of 56 divisions, the Soviet, 114.

The situation in the air over the European fronts is similar: by the fullest count, the Soviet Union alone could muster 4,700 fighters, fighter-bombers, and interceptors, without reinforcement from other theaters; the Western air forces in Europe hold a total of 3,045, of which not more than two-thirds can be considered modern, including all the 594 American fighters and fighter-bombers.

To consider the military balance in the Persian Gulf, with Iran as the possible theater of war, no computation is even needed: against a maximum of four or five American divisions that could eventually be deployed with great difficulties and greater risk, the Soviet Union could send 20 with great ease.

On the last of the "continental" fronts, which cuts across the peninsula of Korea—where sudden war is all too possible, but where a large Soviet intervention now seems most unlikely—it is the Korean forces on both sides that now make the balance. But should Moscow choose to do so, it could add much more to the North Korean strength than the United States could add to that of South Korea.

Thus on every possible major front we encounter the powerful arithmetic of the Soviet army. By integrating reserves with active units and providing full equipment, the Soviet army is a very effective producer of armor-mechanized divisions. Not at all suited for overseas expeditions, dependent on rail transport for large movements between the different fronts separated by several thousand miles, these divisions are nevertheless powerful instruments of offensive war wherever the Soviet Union may seek to enlarge its empire.

With Western air power now offset to a large degree by Soviet air defenses, and with naval power only relevant in the less critical theaters remote from Europe, the Middle East, and East Asia, the ground forces are the basic currency of East-West strategy. Because of the Soviet Union's energetic countering efforts, its advantage in ground forces can no longer be offset by Western strengths in other forms of military power, including (as we shall see later) nuclear.

The combat value of the mass of the Soviet armed forces remains untested by the terrible urgencies of war. But it is possible to estimate their organizational, operational, and tactical competence—if not their fighting spirit—by observing exercises, which show quite clearly that the Soviet armed forces can now execute complicated military operations on a very large scale.

Specifically, we know that the Soviet army can assemble, supply, and send out its long columns of armor and considerable artillery to defeat enemy fronts, not in a steamroller action of costly head-on attacks, as in the past, but rather in quick probes—to find gaps and weak sectors, and to follow with fast-paced penetrations into the rear to achieve great encirclements; to overrun forward air bases, depots, and command centers; to "hug" cities so that tactical nuclear attacks against Soviet advancing forces would hit allied population centers (and would thus be inhibited); and to seize large extents of territory in so doing.

At the same time, raiding forces of the airborne divisions, of the special helicopter-assault brigades, of the "diversionary" and commando units of both military and civilian intelligence can fly into, parachute into, or infiltrate the deep rear in order to seize nuclear-weapons storage sites, attack headquarters and communications centers, sabotage aircraft in their hangars and fire across crowded runways, ambush road convoys, and spread havoc by their mere presence—and by the inevitable tide of false reports about their doings and undoings.

We know that the Soviet air force has enough aircraft, enough bases, and enough quality in men and machines to deny air supremacy to whatever Western air forces it might meet in Europe, the Persian Gulf, or East Asia. Its fighter-interceptors, along with their anti-aircraft defenses, could keep Western air forces from doing much harm to the Soviet army; its long-range strike fighters could reach and bomb Western airfields even in the deepest rear, and its fighter-bombers and ground-attack aircraft could disrupt if not seriously reduce Western ground forces. In theory, Western air forces could eventually prevail in the contest for air supremacy—if the Soviet ground forces had not by then overrun their airfields. One thing is certain: Western air power can give little help to the ground forces in the first days of a war—precisely when air support would be needed most urgently.

We know that the Soviet navy can send out its aircraft, group its ships, and deploy its attack submarines in a concerted worldwide action to stage simultaneous missile strikes on American carrier task forces at sea, certainly in the Indian Ocean, eastern Mediterranean, and northeast Pacific, and possibly in the Atlantic and eastern Pacific as well. Though lacking the floating air power that remains the costly centerpiece of the American navy, the Soviet Union can nevertheless soberly estimate that *if* it attacked first, it could destroy the main fighting strength of the American navy actually at sea. In any event, Soviet attack submarines would endanger the sea connection between the United States and American forces overseas.

So far, not a word has been said about the entire subject of Soviet nuclear weapons. This separation and implied downgrading of the matter corresponds to the strategic logic of the Soviet position against the West. Moscow's protestations of reluctance to use nuclear weapons against the West (China is another matter) may be perfectly sincere. Just as the invader is always peaceful—for he seeks only to advance and not to fight, while it is his victim who causes war by resisting—so the Soviet Union has every reason to avoid nuclear war, because it is now stronger than the West in non-nuclear military forces. Fully able to invade Europe, Iran, or Korean without having to use nuclear weapons, the Soviet Union now needs its nuclear weapons mainly to neutralize the nuclear deterrence of the United States, Britain, and France. Just as it is always the victim who must make war to resist aggression, so the West must rely on the fear of nuclear war to obtain security, by threatening nuclear attacks against invading Soviet forces if they cannot be stopped by non-nuclear means.

To deter such "tactical" attacks—that is, to inhibit the first level of the Western nuclear deterrent in order to restore the full value of its armies for intimidation or actual invasion—the Soviet Union has built up its own "tactical" nuclear forces, in the form of artillery shells, rockets, short-range missiles, and bombs for fighter-class aircraft and strike bombers. The Soviet Union can therefore reply in kind should its invading armies have their victories spoiled by nuclear attacks.

If the West begins to strike at invading Soviet columns with tactical nuclear weapons, the Soviet Union can, in a simple military calculation, use its own tactical nuclear weapons to blast open paths through the alliance front, so that even badly reduced and shocked invasion columns can continue to advance, eventually to reach and "hug" the cities—thereby forcing the alliance to stop its nuclear attacks. In the far more meaningful political calculation, the mere existence of large and very powerful Soviet tactical nuclear forces should inhibit to some extent any Western use of the same weapons.

But the alliance has a most significant advantage that arises from its purely defensive character: at this first level, the entire onus of beginning a war rests on the Soviet Union; it is by *its* decision that the movement of the armies would begin; it is by *its* decision that the invasion of Western territory would continue so that tactical nuclear weapons would be used against its forces, raising the conflict to the second level. Hence the Soviet tactical nuclear forces are not sufficient to dissuade Western use of the same weapons. The Soviets could only use them to achieve physical results (blasting gaps through the front) that would not begin to remedy the catastrophic deterioration of their position from a successful non-nuclear invasion to a nuclear conflict in which no good result could be achieved.

Therefore, to inhibit Western tactical nuclear forces much more powerfully, the Soviet Union maintains another category of nuclear weapons of longer ("intermediate") range, which threaten the cities of

Europe, as well as large military targets in the deep rear, such as air bases. At present, the celebrated SS-20 ballistic missile is the main weapon in this category, which also includes Soviet strike aircraft such as the Su-24 ("Fencer"). With these weapons the interaction between Soviet and Western military power reaches its third level.

Of late, the alliance has begun to deploy intermediate-range cruise and Pershing-2 missiles in Britain, West Germany, and Italy (more are to be deployed in Belgium and Holland). Because they are widely regarded as an entirely different category of weapons, they *are* different politically: the huge controversy surrounding their deployment may enable the cruise and pershing-2 missiles to have a counter-intimidation impact since public opinion views them as an answer to the SS-20s. To that extent, they are *politically* distinct from the far more abundant aircraft bombs and all the other nuclear weapons officially described as tactical. In addition, the new missiles may be more reliable in reaching their targets than strike aircraft with nuclear bombs.

But *strategically* the cruise and Pershing-2 missile are *not* different from the tactical nuclear weapons of the alliance: they too serve to neutralize the non-nuclear strength of the Soviet army, and they too are neutralized in turn by the Soviet nuclear counterthreat against the cities of the alliance. As a matter of physical fact, the cruise and Pershing-2 missiles do not threaten anything not already threatened by alliance weapons classified as tactical; specifically, they do not threaten Soviet cities any more than the tactical nuclear bombs of longer-ranged alliance strike aircraft. Both those aircraft and the new missiles could reach cities in the western part of the Soviet Union; neither is meant to be used against those cities; for both, the relevant targets are Soviet military forces and their bases and command centers.

To neutralize the Soviet third-level threat against the alliance cities in Europe, the new missiles would have to counterthreaten Soviet cities with an equal certainty of complete destruction; because of their vulnerability and range limits, the new missiles cannot do that. Hence the new missiles cannot take the strategic interaction to a fourth level, where the Soviet invasion potential is once again neutralized. The third level thus leaves the Soviet Union in control of the situation, because with or without the cruise and Pershing-2 missiles, the alliance can protect its frontal defenses only at the risk of provoking Soviet nuclear attacks against the cities that those same frontal forces are supposed to protect.

It takes a fourth level to restore a war-avoiding balance, in which this Soviet third-level nuclear threat is itself deterred by American intercontinental nuclear forces capable of inflicting catastrophic destruction on the Soviet Union. Then the Western "tactical" nuclear forces can once again deter a Soviet (non-nuclear) invasion, and the Soviet Union's invasion potential yields neither war options nor the power to intimidate the European allies of the United States.

The Soviet response would be to seek a fifth level of strategic interaction, where the American deterrent would be neutralized by the threat of destroying the intercontinental nuclear force if any were used against Soviet military forces, and American cities if any Soviet cities were destroyed. If the United States government would withdraw its threat of a nuclear attack on the Soviet Union in response to a Soviet attack on European cities, or if American intercontinental nuclear forces could not plausibly threaten the Soviet Union, the strategic interaction would revert to the third level, in which the Soviet threat against the cities of the allies inhibits the West from using its tactical nuclear forces, thus making the world safe for the Soviet army.

One hears it said endlessly that the competition between American and Soviet intercontinental nuclear forces is not only costly and dangerous but also futile, because each side can already destroy the population of the other "many times over." That, however, is a vulgar misunderstanding. It is not to destroy the few hundred cities and larger towns of each side—easy targets neither protected nor concealed—that intercontinental nuclear forces continue to be developed. The purpose is not to threaten cities and towns already abundantly threatened, to "overkill" populations, but rather to threaten the intercontinental nuclear forces themselves: the missiles in their fortified housings, the bomber bases and missile-submarine ports, and the centers of military command and communication for all those forces. Thus there are several thousand targets, as opposed to a few hundred cities and towns, and many of those targets can only be destroyed by very accurate warheads.

At the fifth level of interaction, each side strives to reduce the nuclear-attack strength of the other, by defenses when possible (notably anti-aircraft, against the bombers), but mainly by offensive weapons accurate enough to destroy the weapons of the other side. And it is not enough to be able to threaten the destruction of the weapons: to make the threat effective it is also necessary to demonstrate the ability to destroy them without at the same time destroying the nearby population centers. For if that happens, then all the strategy and all rational purposes come to an end, as the victim will respond by launching his surviving weapons (there will always be some, perhaps many) against the cities of the attacker. For the United States, the competition at this point is driven by the goal of keeping the strategic interaction at the fourth level, where the Soviet army stands deterred; for the Soviet Union, the goal is to reach the fifth level, where American nuclear deterrence is itself deterred.

Because of the goals now pursued, contrary to widespread belief, intercontinental nuclear weapons are steadily becoming *less* destructive in gross explosive power. The goal of each side is to make its forces more accurate and more controllable so that they can destroy small and well-protected targets, and no more. During the 1960s, the United States was still producing weapons of 5 and 9 megatons, while the Soviet Union was producing 20-megaton warheads; nowadays, most new American warheads have yields of less than half a megaton, while most Soviet warheads are below one megaton. As new weapons replace old, the total destructive power of the two intercontinental nuclear arsenals is steadily declining. (A "freeze," incidentally, would put an end to that process.)

The state of the American-Soviet intercontinental nuclear balance is the basic index, the Dow-Jones, of world politics. Directly or through sometimes subtle hopes and fears it shapes much of what American and Soviet leaders feel free to do in world affairs. Two things are quite obvious about the current intercontinental nuclear balance. Both sides can easily destroy the cities and larger towns of the other. And neither can launch an all-out strike that would fully disarm the other's weapons. The competition is now between these two extremes, as each side seeks to protect as many of its weapons as possible while threatening the other's weapons.

Although the United States is by no means inferior across the board, category by category, it is impossible to extract an optimistic estimate from the numbers. There are 1,398 Soviet intercontinental ballistic missiles in underground housings, as against 1,000 American Minuteman missiles; some of the latter have been modernized and others have not, but the Soviet missiles are much larger, with many more warheads (almost 6,000 versus 2,100), which are no longer less accurate than their American counterparts (as was the case till quite recently). No expert disputes the accuracy and reliability of the more modern Soviet ballistic missiles—the four-warhead SS-17 (150 in service), the huge SS-18, with as many as ten warheads (308 in service), and the slightly less modern but more abundant six-warhead SS-19 (330 in service). The combined Soviet force clearly outmatches the 450 one-warhead Minuteman 2s and 550 three-warhead Minuteman 3s. Specifically, the Soviet land-based missiles could now destroy all but a fraction of their

American counterparts, while the latter could not hope to do the same to the Soviet force.

The remaining defect of the Soviet land-based ballistic-missile force is that its warheads are not yet small enough to make the threat of a "clean" disarming strike believable. (The smallest warheads are of half-megaton size, and some are almost a full megaton). The Soviet Union is now developing an entire new group of land-based ballistic missiles: undoubtedly they will be more accurate, and their warheads will be smaller. The new American MX missiles now in production are also meant to be more accurate, although their original purpose was greater survivability, which is dubious, since they will be placed inside fixed housings, though they were built for mobility.

The Soviet force of submarine-carried ballistic missiles is also much larger than the American, with 980 missiles as opposed to 640, but the quality difference is still so great that the American force remains superior. In the first place, most Soviet submarine-lauched missiles are still one-warhead weapons, while their American counterparts have multiple warheads. As a result, the Soviet Union has fewer than 1,000 separate warheads in its submarine force, as against more than 5,000 (much smaller) American warheads.

A greater defect in a force that is, or should be, the ultimate strategic reserve, the best-protected of all in the intercontinental category, is the fact that all the Soviet submarines, except perhaps the very latest, are much noisier and thus more easily detected than their American counterparts.[1] This is all the more striking because the Soviet submarines are much newer on average: between 1974 and 1984 the Soviet Union built 35 Delta-class submarines and one huge Typhoon, as against just four bigger still Ohio-class submarines built by the United States. On the other hand, the latest Soviet submarine-launched missile, the SS-N-20, has such a long range (8,300 kilometers) that it can reach most targets without requiring the submarine carrying it to leave safe waters near the Soviet northern coasts.

Throughout the long years of strategic competition, the Soviet force of intercontinental bombers remained much smaller, and its aircraft much inferior, though this may be about to change. In the latest count, the 297 American bombers, mostly ancient but much modernized B-52s, can be compared to a total of 273 Soviet bombers, including roughly 130 Backfires that are modern and supersonic, but not quite sufficient in range (5,500 kilometers), as well as a greater number of Tu-95s, an aircraft as old as the B-52 but much inferior in every way. Only recently has the Soviet Union started producing a true modern intercontinental bomber, the Blackjack, which is extremely similar to the new American B-15, and is destined to be electronically less advanced but also much faster.

For all their technical inferiority, Soviet bombers are still formidable, simply because the United States has very weak air defenses. While American bombers would have to contend with 1,250 Soviet interceptor-fighters and almost 10,000 anti-aircraft missiles to reach their targets, Soviet bombers would virtually have a free ride against 90 air force and 180 Air National Guard interceptors, and not a single missile.

The Soviet Union destruction of the Korean airliner (flight KAL 007), on September 1, 1983, has been interpreted by some as proof of the incompetence of Soviet territorial air defenses. In one version, which assumes that the attack was made in error, the Soviet radar network is judged grossly incompetent for having failed to distinguish between KAL 007 and the very much smaller RC-135 electronic-reconnaissance aircraft (supposedly the intended victim). In a second version the mere fact that KAL 007 was not actually shot down until two-and-one-half hours after it first entered Soviet airspace over the Kamchatka peninsula is treated as a failure of the system, regardless of whether the aircraft was correctly identified.

These interpretations illustrate very well the difficulty of making operational judgments in a vacuum; the mere fact that the Korean airliner *was* found and reached by a Soviet fighter, that a missile was launched correctly, that it detonated and destroyed a large aircraft is simply taken for granted, as if these were easy things. And indeed they should be for any air-defense system at war, operating day in and day out, with all the habits of combat operations. But on September 1, 1983, until KAL 007 arrived on the scene, Soviet air defenses were at peace, as they have been for almost forty years. To monitor the air space closely, to have the fighters ready at the end of the runway, to have pilots find the target, to have missiles fully operational—to have all this when the action suddenly starts after decades of inaction is not easy at all. The interception of KAL 007 should be compared to the noninterception by American air defenses of more than one Cuban airliner that violated U.S. flight corridors on the Havana-Mexico City route.

Even a delay of two-and-one-half hours would not be significant. But as it happens, the delay was nowhere near so long; KAL 007 first penetrated and then left Soviet airspace (over Kamchatka), before reentering Soviet airspace (near Sakhalin). Its first penetration was very brief, a matter of minutes and forgivable even by the Soviet Union. Its second led to its destruction in short order.

The misidentification theory takes for granted that a 747 can very easily be distinguished from an RC-135. That is simply not the case; identification depends, among other things (size, aspect, frequencies, type of radar and displays), on atmospheric conditions. But as it happens, it is certain that the Soviet air-defense controllers knew exactly what they were destroying; this is one case where the negative evidence prevails. As in the Sherlock Holmes story, the dog that did not bark is definite proof: though Soviet air-defense controllers could have confused the KAL 007 radar image with that of an RC-135, the scheduled Korean Air Lines flight from Anchorage, Alaska, to Seoul, Korea, which Soviet radar would routinely track, had to be *somewhere* on the radar screens. If it was not, only two possibilities remained— either that KAL 007 had crashed into the sea without any signal at all, or else that the aircraft being intercepted was in fact KAL 007. So to believe in the misidentification theory, we have to assume that Soviet air-defense controllers not only confused the two radar images but believed that KAL 007 had mysteriously fallen out of the sky without even a few seconds in which to transmit a "mayday" call. Thus once again we must resist the seductive urge to believe in Soviet ineptness.

The Soviet Union continues to make a large effort in strategic defense, maintaining costly forces to fight what can be fought (the bombers and cruise missiles), doing all it can to develop anti-ballistic-missile defenses, and keeping up a nationwide civil-defense program combining highly realistic with merely symbolic arrangements, from shelters to evacuation. The United States by contrast is pursuing innovation in offensive weapons (cruise missiles for the bombers, surface ships, and attack submarines, and Trident-2 submarine-launched missiles) and exploring many highly advanced defensive schemes based on satellite-mounted weapons, but it has no serious civil defense.

One could add details and nuances to the estimate of the Soviet Union's intercontinental nuclear strength and homeland defenses, but the result would not change, for the two forces do not have the same task. The United States must rely on believable threats to use its intercontinental nuclear forces to offset the Soviet Union's non-nuclear superiority and "tactical" nuclear parity. Otherwise matters would stand at the third level, where there is nothing to stop Soviet military intimidation of America's allies—who then could scarcely remain allies. The Soviet Union by contrast need only make the American intercontinental nuclear threat unbelievable in order to recover the invasion potential of its armies, thus restoring their power to intimidate or actually invade. To do that the Soviet Union does not even need intercontinental nuclear superiority, which it is striving so hard to achieve. But the United States does need a margin of intercontinental nuclear strength merely to keep the overall military balance duly balanced.

Hence "parity" (shorthand for strategic-nuclear parity) is or should be fundamentally unacceptable to the United States. Any true parity between the intercontinental nuclear forces of each side must leave the United States militarily inferior in all the continental theaters where the Soviet army can muster its power—namely Europe, Iran (and thus the Persian Gulf), and Korea. And that is the situation that now prevails, the true cause of today's anxieties for world peace.

Note

1. Among the 80 Soviet ballistic-missile submarines in service, some 22 are ancient diesel and early vintage nuclear boats that have every right to be noisy, but these account for fewer than 60 of the 980 missile tubes. The bulk of the force should be much less noisy than it is, raising some interesting questions about Soviet design, or perhaps strategy.

INDEX